Cambridge Tracts in Mathematics and Mathematical Physics

GENERAL EDITORS
P. HALL, F.R.S. AND F. SMITHIES, PH.D.

No. 2

The Integration of Functions of a Single Variable

THE
INTEGRATION OF FUNCTIONS
OF A SINGLE VARIABLE

BY THE LATE

G. H. HARDY

SECOND EDITION

CAMBRIDGE
AT THE UNIVERSITY PRESS
1966

CAMBRIDGE UNIVERSITY PRESS
Cambridge, New York, Melbourne, Madrid, Cape Town, Singapore,
São Paulo, Delhi, Dubai, Tokyo, Mexico City

Cambridge University Press
The Edinburgh Building, Cambridge CB2 8RU, UK

Published in the United States of America by Cambridge University Press, New York

www.cambridge.org
Information on this title: www.cambridge.org/9780521172226

First edition 1905
Second edition 1916
Reprinted 1928, 1958, 1966
First paperback edition 2010

A catalogue record for this publication is available from the British Library

ISBN 978-0-521-05205-4 Hardback
ISBN 978-0-521-17222-6 Paperback

PREFACE

THIS tract has been long out of print, and there is still some demand for it. I did not publish a second edition before, because I intended to incorporate its contents in a larger treatise on the subject which I had arranged to write in collaboration with Dr Bromwich. Four or five years have passed, and it seems very doubtful whether either of us will ever find the time to carry out our intention. I have therefore decided to republish the tract.

The new edition differs from the first in one important point only. In the first edition I reproduced a proof of Abel's which Mr J. E. Littlewood afterwards discovered to be invalid. The correction of this error has led me to rewrite a few sections (pp. 36–41 of the present edition) completely. The proof which I give now is due to Mr H. T. J. Norton. I am also indebted to Mr Norton, and to Mr S. Pollard, for many other criticisms of a less important character.

G. H. H.

January 1916.

CONTENTS

PAGE

I. Introduction 1

II. Elementary functions and their classification 3

III. The integration of elementary functions. Summary of results . 8

IV. The integration of rational functions 12

   1-3. The method of partial fractions 12

     4. Hermite's method of integration 15

     5. Particular problems of integration 17

     6. The limitations of the methods of integration . . . 19

     7. Conclusion 21

V. The integration of algebraical functions 22

     1. Algebraical functions 22

     2. Integration by rationalisation. Integrals associated with
        conics 23

   3-6. The integral $\int R\{x, \sqrt{(ax^2 + 2bx + c)}\}\,dx$ 24

     7. Unicursal plane curves 30

     8. Particular cases 33

     9. Unicursal curves in space 35

   10. Integrals of algebraical functions in general . . . 35

 11-14. The general form of the integral of an algebraical function.
        Integrals which are themselves algebraical . . . 36

   15. Discussion of a particular case 42

   16. The transcendence of e^x and $\log x$ 44

   17. Laplace's principle 44

   18. The general form of the integral of an algebraical function
        (*continued*). Integrals expressible by algebraical functions
        and logarithms 45

                                                                                PAGE
    19. Elliptic and pseudo-elliptic integrals. Binomial integrals   47

    20. Curves of deficiency 1. The plane cubic   48

    21. Degenerate Abelian integrals   50

    22. The classification of elliptic integrals   51

VI. The integration of transcendental functions   52

    1. Preliminary   52

    2. The integral $\int R\left(e^{ax},\ e^{bx},\ \ldots,\ e^{kx}\right)dx$ 52

    3. The integral $\int P\left(x,\ e^{ax},\ e^{bx},\ \ldots\right)dx$ 55

    4. The integral $\int e^{x}\,R\left(x\right)dx$. The logarithm-integral   .   .   56

    5. Liouville's general theorem   59

    6. The integral $\int \log x\,R\left(x\right)dx$ 60

    7. Conclusion   62

    Appendix I.   Bibliography   63

    Appendix II.   On Abel's proof of the theorem of v., § 11   .   66

THE INTEGRATION OF FUNCTIONS OF A SINGLE VARIABLE

I. Introduction

The problem considered in the following pages is what is sometimes called the problem of 'indefinite integration' or of 'finding a function whose differential coefficient is a given function'. These descriptions are vague and in some ways misleading; and it is necessary to define our problem more precisely before we proceed further.

Let us suppose for the moment that $f(x)$ is a real continuous function of the real variable x. We wish to determine a function y whose differential coefficient is $f(x)$, or to solve the equation

$$\frac{dy}{dx} = f(x) \quad \dots\dots\dots\dots\dots\dots\dots\dots\dots(1).$$

A little reflection shows that this problem may be analysed into a number of parts.

We wish, first, to know whether such a function as y necessarily exists, whether the equation (1) has always a solution; whether the solution, if it exists, is unique; and what relations hold between different solutions, if there are more than one. The answers to these questions are contained in that part of the theory of functions of a real variable which deals with 'definite integrals'. The definite integral

$$y = \int_a^x f(t)\, dt \quad \dots\dots\dots\dots\dots\dots\dots(2),$$

which is defined as the limit of a certain sum, is a solution of the equation (1). Further

$$y + C \dots\dots\dots\dots\dots\dots\dots\dots\dots(3),$$

where C is an arbitrary constant, is also a solution, and all solutions of (1) are of the form (3).

These results we shall take for granted. The questions with which we shall be concerned are of a quite different character. They are questions as to the functional form of y when $f(x)$ is a function of some stated form. It is sometimes said that the problem of indefinite integration is that of 'finding an actual expression for y when $f(x)$ is given'. This statement is however still lacking in precision. The theory of definite integrals provides us not only with a proof of the existence of a solution, but also with an expression for it, an expression in the form of a limit. The problem of indefinite integration can be stated precisely only when we introduce sweeping restrictions as to the classes of functions and the modes of expression which we are considering.

Let us suppose that $f(x)$ belongs to some special class of functions $\mathfrak{F}$. Then we may ask whether y is itself a member of $\mathfrak{F}$, or can be expressed, according to some simple standard mode of expression, in terms of functions which are members of $\mathfrak{F}$. To take a trivial example, we might suppose that $\mathfrak{F}$ is the class of polynomials with rational coefficients: the answer would then be that y is in all cases itself a member of $\mathfrak{F}$.

The range and difficulty of our problem will depend upon our choice of (1) a class of functions and (2) a standard 'mode of expression'. We shall, for the purposes of this tract, take $\mathfrak{F}$ to be the class of *elementary functions*, a class which will be defined precisely in the next section, and our mode of expression to be that of *explicit expression in finite terms*, *i.e.* by formulae which do not involve passages to a limit.

One or two more preliminary remarks are needed. The subject-matter of the tract forms a chapter in the 'integral calculus'[*], but does not depend in any way on any direct theory of integration. Such an equation as

$$y = \int f(x)\,dx \dots\dots\dots\dots\dots\dots(4)$$

is to be regarded as merely another way of writing (1): the integral sign is used merely on grounds of technical convenience, and might be eliminated throughout without any substantial change in the argument.

[*] Euler, the first systematic writer on the 'integral calculus', defined it in a manner which identifies it with the theory of differential equations: 'calculus integralis est methodus, ex data differentialium relatione inveniendi relationem ipsarum quantitatum' (*Institutiones calculi integralis*, p. 1). We are concerned only with the special equation (1), but all the remarks we have made may be generalised so as to apply to the wider theory.

The variable x is in general supposed to be complex. But the tract should be intelligible to a reader who is not acquainted with the theory of analytic functions and who regards x as real and the functions of x which occur as real or complex functions of a real variable.

The functions with which we shall be dealing will always be such as are regular except for certain special values of x. These values of x we shall simply ignore. The meaning of such an equation as

$$\int \frac{dx}{x} = \log x$$

is in no way affected by the fact that $1/x$ and $\log x$ have infinities for $x = 0$.

II. Elementary functions and their classification

An *elementary function* is a member of the class of functions which comprises

(i)   rational functions,

(ii)   algebraical functions, explicit or implicit,

(iii)   the exponential function e^x,

(iv)   the logarithmic function $\log x$,

(v)   all functions which can be defined by means of any finite combination of the symbols proper to the preceding four classes of functions.

A few remarks and examples may help to elucidate this definition.

1. A *rational function* is a function defined by means of any finite combination of the elementary operations of addition, multiplication, and division, operating on the variable x.

It is shown in elementary algebra that any rational function of x may be expressed in the form

$$f(x) = \frac{a_0 x^m + a_1 x^{m-1} + \dots + a_m}{b_0 x^n + b_1 x^{n-1} + \dots + b_n},$$

where m and n are positive integers, the a's and b's are constants, and the numerator and denominator have no common factor. We shall adopt this expression as the standard form of a rational function. It is hardly necessary to remark that it is in no way involved in the definition of a rational function that these constants should be rational or algebraical* or real *numbers*. Thus

$$\frac{x^2 + x + i\sqrt{2}}{x\sqrt{2} - e}$$

is a rational function.

2. An *explicit algebraical function* is a function defined by means of any finite combination of the four elementary operations and any finite number of operations of root extraction. Thus

$$\frac{\sqrt{(1+x)} - \sqrt[3]{(1-x)}}{\sqrt{(1+x)} + \sqrt[3]{(1-x)}}, \quad \sqrt{\{x + \sqrt{(x + \sqrt{x})}\}}, \quad \left(\frac{x^2 + x + i\sqrt{2}}{x\sqrt{2} - e}\right)^{\frac{2}{3}}$$

are explicit algebraical functions. And so is $x^{m/n}$ (*i.e.* $\sqrt[n]{x^m}$) for any integral values of m and n. On the other hand

$$x^{\sqrt{2}}, \quad x^{1+i}$$

are not algebraical functions at all, but transcendental functions, as irrational or complex powers are defined by the aid of exponentials and logarithms.

Any explicit algebraical function of x satisfies an equation

$$P_0 y^n + P_1 y^{n-1} + \dots + P_n = 0$$

whose coefficients are polynomials in x. Thus, for example, the function

$$y = \sqrt{x} + \sqrt{(x + \sqrt{x})}$$

satisfies the equation

$$y^4 - (4y^2 + 4y + 1)\,x = 0.$$

The converse is not true, since it has been proved that in general equations of degree higher than the fourth have no roots which are explicit algebraical functions of their coefficients. A simple example is given by the equation

$$y^5 - y - x = 0.$$

We are thus led to consider a more general class of functions, *implicit* algebraical functions, which includes the class of explicit algebraical functions.

* An algebraical number is a number which is the root of an algebraical equation whose coefficients are integral. It is known that there are numbers (such as e and π) which are not roots of any such equation. See, for example, Hobson's *Squaring the circle* (Cambridge, 1913).

3. An *algebraical function* of x is a function which satisfies an equation

$$P_0 y^n + P_1 y^{n-1} + \dots + P_n = 0 \quad \dots\dots\dots\dots\dots(1)$$

whose coefficients are polynomials in x.

Let us denote by $P(x, y)$ a polynomial such as occurs on the left-hand side of (1). Then there are two possibilities as regards any particular polynomial $P(x, y)$. Either it is possible to express $P(x, y)$ as the product of two polynomials of the same type, neither of which is a mere constant, or it is not. In the first case $P(x, y)$ is said to be *reducible*, in the second *irreducible*. Thus

$$y^4 - x^2 = (y^2 + x)(y^2 - x)$$

is reducible, while both $y^2 + x$ and $y^2 - x$ are irreducible.

The equation (1) is said to be reducible or irreducible according as its left-hand side is reducible or irreducible. A reducible equation can always be replaced by the logical alternative of a number of irreducible equations. Reducible equations are therefore of subsidiary importance only ; and we shall always suppose that the equation (1) is irreducible.

An algebraical function of x is regular except at a finite number of points which are *poles* or *branch points* of the function. Let D be any closed simply connected domain in the plane of x which does not include any branch point. Then there are n and only n distinct functions which are one-valued in D and satisfy the equation (1). These n functions will be called the *roots* of (1) in D. Thus if we write

$$x = r(\cos\theta + i\sin\theta),$$

where $-\pi < \theta \leqslant \pi$, then the roots of

$$y^2 - x = 0,$$

in the domain

$$0 < r_1 \leqslant r \leqslant r_2, \quad -\pi < -\pi + \delta \leqslant \theta \leqslant \pi - \delta < \pi,$$

are $\sqrt{x}$ and $-\sqrt{x}$, where

$$\sqrt{x} = \sqrt{r}(\cos\tfrac{1}{2}\theta + i\sin\tfrac{1}{2}\theta).$$

The relations which hold between the different roots of (1) are of the greatest importance in the theory of functions*. For our present purposes we require only the two which follow.

(i) Any symmetric polynomial in the roots $y_1, y_2, \dots, y_n$ of (1) is a rational function of x.

---

* For fuller information the reader may be referred to Appell and Goursat's *Théorie des fonctions algébriques*.

(ii) Any symmetric polynomial in $y_2, y_3, \ldots, y_n$ is a polynomial in y_1 with coefficients which are rational functions of x.

The first proposition follows directly from the equations

$$\Sigma\, y_1 y_2 \ldots y_s = (-1)^s (P_{n-s}/P_0) \qquad (s = 1, 2, \ldots, n).$$

To prove the second we observe that

$$\underset{2,3,\ldots}{\Sigma}\ y_2 y_3 \ldots y_s = \underset{1,2,\ldots}{\Sigma}\ y_1 y_2 \ldots y_{s-1} - y_1 \underset{2,3,\ldots}{\Sigma}\ y_2 y_3 \ldots y_{s-1}.$$

so that the theorem is true for $\Sigma\, y_2 y_3 \ldots y_s$ if it is true for $\Sigma\, y_2 y_3 \ldots y_{s-1}$. It is certainly true for

$$y_2 + y_3 + \ldots + y_n = (y_1 + y_2 + \ldots + y_n) - y_1.$$

It is therefore true for $\Sigma\, y_2 y_3 \ldots y_s$, and so for any symmetric polynomial in $y_2, y_3, \ldots, y_n$.

4. Elementary functions which are not rational or algebraical are called *elementary transcendental functions* or elementary transcendents. They include all the remaining functions which are of ordinary occurrence in elementary analysis.

The trigonometrical (or circular) and hyperbolic functions, direct and inverse, may all be expressed in terms of exponential or logarithmic functions by means of the ordinary formulae of elementary trigonometry. Thus, for example,

$$\sin x = \frac{e^{ix} - e^{-ix}}{2i}, \qquad\qquad \sinh x = \frac{e^x - e^{-x}}{2},$$

$$\text{arc tan } x = \frac{1}{2i} \log \left(\frac{1 + ix}{1 - ix} \right), \qquad \text{arg tanh } x = \frac{1}{2} \log \left(\frac{1+x}{1-x} \right).$$

There was therefore no need to specify them particularly in our definition.

The elementary transcendents have been further classified in a manner first indicated by Liouville*. According to him a function is a transcendent *of the first order* if the signs of exponentiation or of the taking of logarithms which occur in the formula which defines it apply only to rational or algebraical functions. For example

$$xe^{-x^2}, \quad e^{x^2} + e^x \sqrt{(\log x)}$$

are of the first order ; and so is

$$\text{arc tan } \frac{y}{\sqrt{(1 + x^2)}},$$

---

* 'Mémoire sur la classification des transcendantes, et sur l'impossibilité d'exprimer les racines de certaines équations en fonction finie explicite des coefficients', *Journal de mathématiques*, ser. 1, vol. 2, 1837, pp. 56–104 ; 'Suite du mémoire...', *ibid.* vol. 3, 1838, pp. 523–546.

where y is defined by the equation

$$y^5 - y - x = 0 \; ;$$

and so is the function y defined by the equation

$$y^5 - y - e^x \log x = 0.$$

An elementary transcendent *of the second order* is one defined by a formula in which the exponentiations and takings of logarithms are applied to rational or algebraical functions or to transcendents of the first order. This class of functions includes many of great interest and importance, of which the simplest are

$$e^{e^x}, \; \log \log x.$$

It also includes irrational and complex powers of x, since, *e.g.*,

$$x^{\surd 2} = e^{\surd 2 \log x}, \quad x^{1+i} = e^{(1+i)\log x} \; ;$$

the function $x^x = e^{x \log x} \; ;$

and the logarithms of the circular functions.

It is of course presupposed in the definition of a transcendent of the second kind that the function in question is incapable of expression as one of the first kind or as a rational or algebraical function. The function

$$e^{\log R(x)},$$

where $R(x)$ is rational, is not a transcendent of the second kind, since it can be expressed in the simpler form $R(x)$.

It is obvious that we can in this way proceed to define transcendents of the nth order for all values of n. Thus

$$\log \log \log x, \; \log \log \log \log x, \; \ldots\ldots$$

are of the third, fourth, …… orders.

Of course a similar classification of algebraical functions can be and has been made. Thus we may say that

$$\surd x, \; \surd(x + \surd x), \; \surd\{x + \surd(x + \surd x)\}, \; \ldots\ldots$$

are algebraical functions of the first, second, third, …… orders. But the fact that there is a general theory of algebraical equations and therefore of *implicit* algebraical functions has deprived this classification of most of its importance. There is no such general theory of elementary transcendental equations*, and therefore we shall not

---

* The natural generalisations of the theory of algebraical equations are to be found in parts of the theory of differential equations. See Königsberger, 'Bemerkungen zu Liouville's Classificirung der Transcendenten', *Math. Annalen*, vol. 28, 1886, pp. 483–492.

rank as 'elementary' functions defined by transcendental equations such as

$$y = x \log y,$$

but incapable (as Liouville has shown that in this case y is incapable) of explicit expression in finite terms.

5. The preceding analysis of elementary transcendental functions rests on the following theorems :

(a) e^x is not an algebraical function of x ;

(b) $\log x$ is not an algebraical function of x ;

(c) $\log x$ is not expressible in finite terms by means of signs of exponentiation and of algebraical operations, explicit or implicit* ;

(d)   transcendental functions of the first, second, third, ... orders actually exist.

A proof of the first two theorems will be given later, but limitations of space will prevent us from giving detailed proofs of the third and fourth.   Liouville has given interesting extensions of some of these theorems : he has proved, for example, that no equation of the form

$$A e^{\alpha p} + B e^{\beta p} + \dots + R e^{\rho p} = S,$$

where p, A, B, ..., R, S are algebraical functions of x, and $\alpha, \beta, \dots, \rho$ different constants, can hold for all values of x.

III.   The integration of elementary functions.
Summary of results

In the following pages we shall be concerned exclusively with the problem of the integration of elementary functions.   We shall endeavour to give as complete an account as the space at our disposal permits of the progress which has been made by mathematicians towards the solution of the two following problems :

(i)   *if $f(x)$ is an elementary function, how can we determine whether its integral is also an elementary function?*

(ii)   *if the integral is an elementary function, how can we find it?*

It would be unreasonable to expect complete answers to these questions.   But sufficient has been done to give us a tolerably complete insight into the nature of the answers, and to ensure that it

---

* For example, $\log x$ cannot be equal to e^y, where y is an algebraical function of x.

shall not be difficult to find the complete answers in any particular case which is at all likely to occur in elementary analysis or in its applications.

It will probably be well for us at this point to summarise the principal results which have been obtained.

1. The integral of a rational function (IV.) is *always* an elementary function. It is either rational or the sum of a rational function and of a finite number of constant multiples of logarithms of rational functions (IV., 1).

If certain constants which are the roots of an algebraical equation are treated as known then the form of the integral can always be determined completely. But as the roots of such equations are not in general capable of explicit expression in finite terms, it is not in general possible to express the integral in an absolutely explicit form (IV.; 2, 3).

We can always determine, by means of a finite number of the elementary operations of addition, multiplication, and division, whether the integral is rational or not. If it is rational, we can determine it completely by means of such operations; if not, we can determine its rational part (IV.; 4, 5).

The solution of the problem in the case of rational functions may therefore be said to be complete; for the difficulty with regard to the explicit solution of algebraical equations is one not of inadequate knowledge but of proved impossibility (IV., 6).

2. The integral of an algebraical function (V.), explicit or implicit, may or may not be elementary.

If y is an algebraical function of x then the integral $\int y\,dx$, or, more generally, the integral

$$\int R\,(x,\,y)\,dx,$$

where R denotes a rational function, is, if an elementary function, either algebraical or the sum of an algebraical function and of a finite number of constant multiples of logarithms of algebraical functions. All algebraical functions which occur in the integral are *rational functions of x and y* (V.; 11–14, 18).

These theorems give a precise statement of a general principle enunciated by Laplace* : '*l'intégrale d'une fonction différentielle*

* *Théorie analytique des probabilités*, p. 7.

(*algébrique*) *ne peut contenir d'autres quantités radicaux que celles qui entrent dans cette fonction*'; and, we may add, cannot contain *exponentials* at all. Thus it is impossible that

$$\int \frac{dx}{\sqrt{(1 + x^2)}}$$

should contain e^x or $\sqrt{(1-x)}$: the appearance of these functions in the integral could only be apparent, and they could be eliminated before differentiation. Laplace's principle really rests on the fact, of which it is easy enough to convince oneself by a little reflection and the consideration of a few particular cases (though to give a rigorous proof is of course quite another matter), that *differentiation will not eliminate exponentials or algebraical irrationalities*. Nor, we may add, will it eliminate logarithms except when they occur in the simple form

$$A \log \phi (x),$$

where A is a constant, and this is why logarithms can only occur in this form in the integrals of rational or algebraical functions.

We have thus a general knowledge of the form of the integral of an algebraical function y, when it is itself an elementary function. Whether this is so or not of course depends on the nature of the equation $f(x, y) = 0$ which defines y. If this equation, when interpreted as that of a curve in the plane (x, y), represents a *unicursal* curve, *i.e.* a curve which has the maximum number of double points possible for a curve of its degree, or whose *deficiency* is zero, then x and y can be expressed simultaneously as rational functions of a third variable t, and the integral can be reduced by a substitution to that of a rational function (v.; 2, 7–9). In this case, therefore, the integral is always an/elementary function. But this condition, though sufficient, is not necessary. It is in general true that, when $f(x, y) = 0$ is not unicursal, the integral is not an elementary function but a new transcendent; and we are able to classify these transcendents according to the deficiency of the curve. If, for example, the deficiency is unity, then the integral is in general a transcendent of the kind known as *elliptic integrals*, whose characteristic is that they can be transformed into integrals containing no other irrationality than the square root of a polynomial of the third or fourth degree (v., 20). But there are infinitely many cases in which the integral can be expressed by algebraical functions and logarithms. Similarly there are infinitely many cases in which integrals associated with curves whose deficiency is greater than unity are in reality reducible to elliptic integrals. Such abnormal cases have formed the subject of many exceedingly interesting researches, but no general method has been devised by which we can always tell, after a finite series of operations, whether any given integral is really elementary, or elliptic, or belongs to a higher order of transcendents.

When $f(x, y) = 0$ is unicursal we can carry out the integration completely in exactly the same sense as in the case of rational functions. In particular, if the integral is *algebraical* then it can be found by means of elementary operations which are always practicable. And it has been shown, more generally, that we can always determine by means of such operations whether the integral of any given algebraical function is algebraical or not, and evaluate the integral when it is algebraical. And although the general problem of determining whether any given integral is an elementary function, and calculating it if it is one, has not been solved, the solution in the particular case in which the deficiency of the curve $f(x, y) = 0$ is unity is as complete as it is reasonable to expect any possible solution to be.

3. The theory of the integration of transcendental functions (VI.) is naturally much less complete, and the number of classes of such functions for which general methods of integration exist is very small. These few classes are, however, of extreme importance in applications (VI. ; 2, 3).

There is a general theorem concerning the form of an integral of a transcendental function, when it is itself an elementary function, which is quite analogous to those already stated for rational and algebraical functions. The general statement of this theorem will be found in VI., § 5 ; it shows, for instance, that the integral of a rational function of x, e^x and $\log x$ is either a rational function of those functions or the sum of such a rational function and of a finite number of constant multiples of logarithms of similar functions. From this general theorem may be deduced a number of more precise results concerning integrals of more special forms, such as

$$\int y e^x \, dx, \quad \int y \log x \, dx,$$

where y is an algebraical function of x (VI. ; 4, 6).

IV. Rational functions

1. It is proved in treatises on algebra* that any polynomial

$$Q(x) = b_0 x^n + b_1 x^{n-1} + \ldots + b_n$$

can be expressed in the form

$$b_0 (x-a_1)^{n_1} (x-a_2)^{n_2} \ldots (x-a_r)^{n_r},$$

where $n_1, n_2, \ldots$ are positive integers whose sum is n, and $a_1, a_2, \ldots$ are constants ; and that any rational function $R(x)$, whose denominator is $Q(x)$, may be expressed in the form

$$A_0 x^p + A_1 x^{p-1} + \ldots + A_p + \sum_{s=1}^{r} \left\{ \frac{\beta_{s,1}}{x-a_s} + \frac{\beta_{s,2}}{(x-a_s)^2} + \ldots + \frac{\beta_{s,n_s}}{(x-a_s)^{n_s}} \right\},$$

where $A_0, A_1, \ldots, \beta_{s,1}, \ldots$ are also constants. It follows that

$$\int R(x)\, dx = A_0 \frac{x^{p+1}}{p+1} + A_1 \frac{x^p}{p} + \ldots + A_p x + C$$
$$+ \sum_{s=1}^{r} \left\{ \beta_{s,1} \log (x-a_s) - \frac{\beta_{s,2}}{x-a_s} - \ldots - \frac{\beta_{s,n_s}}{(n_s-1)(x-a_s)^{n_s-1}} \right\}.$$

From this we conclude that *the integral of any rational function is an elementary function which is rational save for the possible presence of logarithms of rational functions.* In particular the integral will be *rational* if each of the numbers $\beta_{s,1}$ is zero: this condition is evidently necessary and sufficient. A necessary but not sufficient condition is that $Q(x)$ should contain no simple factors.

The integral of the general rational function may be expressed in a very simple and elegant form by means of symbols of differentiation. We may suppose for simplicity that the degree of $P(x)$ is less than that of $Q(x)$; this can of course always be ensured by subtracting a polynomial from $R(x)$. Then

$$R(x) = \frac{P(x)}{Q(x)}$$
$$= \frac{1}{(n_1-1)!\,(n_2-1)!\ldots(n_r-1)!} \frac{\partial^{n-r}}{\partial a_1^{n_1-1} \partial a_2^{n_2-1} \ldots \partial a_r^{n_r-1}} \frac{P(x)}{Q_0(x)},$$

where
$$Q_0(x) = b_0 (x-a_1)(x-a_2) \ldots (x-a_r).$$

Now
$$\frac{P(x)}{Q_0(x)} = \varpi_0(x) + \sum_{s=1}^{r} \frac{P(a_s)}{(x-a_s)\,Q_0'(a_s)},$$

---

* See, *e.g.*, Weber's *Traité d'algèbre supérieure* (French translation by J. Griess, Paris, 1898), vol. 1, pp. 61–64, 143–149, 350–353 ; or Chrystal's *Algebra*, vol. 1, pp. 151–162.

where $\varpi_0(x)$ is a polynomial; and so

$$\int R(x)\,dx$$
$$= \frac{1}{(n_1-1)!\dots(n_r-1)!}\frac{\partial^{n-r}}{\partial a_1^{n_1-1}\dots\partial a_r^{n_r-1}}\left[\Pi_0(x)+\sum_{s=1}^{r}\frac{P(a_s)}{Q_0'(a_s)}\log(x-a_s)\right],$$

where $\qquad\qquad \Pi_0(x)=\int\varpi_0(x)\,dx.$

But $\qquad\qquad \Pi(x)=\dfrac{\partial^{n-r}\,\Pi_0(x)}{\partial a_1^{n_1-1}\,\partial a_2^{n_2-1}\dots\partial a_r^{n_r-1}}$

is also a polynomial, and the integral contains no polynomial term, since the degree of $P(x)$ is less than that of $Q(x)$. Thus $\Pi(x)$ must vanish identically, so that

$$\int R(x)\,dx$$
$$= \frac{1}{(n_1-1)!\dots(n_r-1)!}\frac{\partial^{n-r}}{\partial a_1^{n_1-1}\dots\partial a_r^{n_r-1}}\left[\sum_{s=1}^{r}\frac{P(a_s)}{Q_0'(a_s)}\log(x-a_s)\right].$$

For example

$$\int\frac{dx}{\{(x-a)(x-b)\}^2}=\frac{\partial^2}{\partial a\,\partial b}\left\{\frac{1}{a-b}\log\left(\frac{x-a}{x-b}\right)\right\}.$$

That $\Pi_0(x)$ is annihilated by the partial differentiations performed on it may be verified directly as follows. We obtain $\Pi_0(x)$ by picking out from the expansion

$$\frac{P(x)}{x^r}\left(1+\frac{a_1}{x}+\frac{a_1^2}{x^2}+\dots\right)\left(1+\frac{a_2}{x}+\frac{a_2^2}{x^2}+\dots\right)\dots\dots$$

the terms which involve positive powers of x. Any such term is of the form

$$A\,x^{\nu-r-s_1-s_2-\dots}\,a_1^{s_1}\,a_2^{s_2}\dots,$$

where $\qquad\qquad s_1+s_2+\dots\leqslant\nu-r\leqslant m-r,$

m being the degree of P. It follows that

$$s_1+s_2+\dots<n-r=(m_1-1)+(m_2-1)+\dots;$$

so that at least one of $s_1,\ s_2,\dots$ must be less than the corresponding one of $m_1-1,\ m_2-1,\dots$.

It has been assumed above that if

$$F(x,a)=\int f(x,a)\,dx,$$

then $\qquad\qquad \dfrac{\partial F}{\partial a}=\int\dfrac{\partial f}{\partial a}\,dx.$

The first equation means that $f = \dfrac{\partial F}{\partial x}$ and the second that $\dfrac{\partial f}{\partial a} = \dfrac{\partial^2 F}{\partial x \partial a}$. As it follows from the first that $\dfrac{\partial f}{\partial a} = \dfrac{\partial^2 F}{\partial a \partial x}$, what has really been assumed is that

$$\frac{\partial^2 F}{\partial a \partial x} = \frac{\partial^2 F}{\partial x \partial a}.$$

It is known that this equation is always true for $x = x_0$, $a = a_0$ if a circle can be drawn in the plane of (x, a) whose centre is (x_0, a_0) and within which the differential coefficients are continuous.

2. It appears from §1 that the integral of a rational function is in general composed of two parts, one of which is a rational function and the other a function of the form

$$\Sigma A \, \log (x - a) \quad \dots\dots\dots\dots\dots\dots\dots(1).$$

We may call these two functions the *rational part* and the *transcendental part* of the integral. It is evidently of great importance to show that the 'transcendental part' of the integral is really transcendental and cannot be expressed, wholly or in part, as a rational or algebraical function.

We are not yet in a position to prove this completely*; but we can take the first step in this direction by showing that *no sum of the form* (1) *can be rational, unless every A is zero.*

Suppose, if possible, that

$$\Sigma A \, \log (x - a) = \frac{P(x)}{Q(x)} \quad \dots\dots\dots\dots\dots(2),$$

where P and Q are polynomials without common factor. Then

$$\Sigma \frac{A}{x - a} = \frac{P'Q - PQ'}{Q^2} \quad \dots\dots\dots\dots\dots(3).$$

Suppose now that $(x - p)^r$ is a factor of Q. Then $P'Q - PQ'$ is divisible by $(x - p)^{r-1}$ and by no higher power of $x - p$. Thus the right-hand side of (3), when expressed in its lowest terms, has a factor $(x - p)^{r+1}$ in its denominator. On the other hand the left-hand side, when expressed as a rational fraction in its lowest terms, has no repeated factor in its denominator. Hence $r = 0$, and so Q is a constant. We may therefore replace (2) by

$$\Sigma A \, \log (x - a) = P(x),$$

and (3) by

$$\Sigma \frac{A}{x - a} = P'(x).$$

Multiplying by $x - a$, and making x tend to a, we see that $A = 0$.

---

* The proof will be completed in v., 16.

3. The method of § 1 gives a complete solution of the problem if the roots of $Q(x) = 0$ can be determined; and in practice this is usually the case. But this case, though it is the one which occurs most frequently in practice, is from a theoretical point of view an exceedingly special case. The roots of $Q(x) = 0$ are not in general explicit algebraical functions of the coefficients, and cannot as a rule be determined in any explicit form. The method of partial fractions is therefore subject to serious limitations. For example, we cannot determine, by the method of decomposition into partial fractions, such an integral as

$$\int \frac{4x^9 + 21x^6 + 2x^3 - 3x^2 - 3}{(x^7 - x + 1)^2}\, dx,$$

or even determine whether the integral is rational or not, although it is in reality a very simple function. A high degree of importance therefore attaches to the further problem of determining the integral of a given rational function so far as possible in an absolutely explicit form and by means of operations which are always practicable.

It is easy to see that a complete solution of this problem cannot be looked for.

Suppose for example that $P(x)$ reduces to unity, and that $Q(x) = 0$ is an equation of the fifth degree, whose roots $a_1, a_2, \ldots a_5$ are all distinct and not capable of explicit algebraical expression.

Then
$$\int R(x)\, dx = \overset{5}{\underset{1}{\Sigma}}\, \frac{\log (x - a_s)}{Q'(a_s)}$$

$$= \log \overset{5}{\underset{1}{\Pi}}\, \{(x - a_s)^{1/Q'(a_s)}\},$$

and it is only if at least two of the numbers $Q'(a_s)$ are commensurable that any two or more of the factors $(x - a_s)^{1/Q'(a_s)}$ can be associated so as to give a single term of the type $A \log S(x)$, where $S(x)$ is rational. In general this will not be the case, and so it will not be possible to express the integral in any finite form which does not explicitly involve the roots. A more precise result in this connection will be proved later (§ 6).

4. The first and most important part of the problem has been solved by Hermite, who has shown that the *rational part* of the integral can always be determined without a knowledge of the roots of $Q(x)$, and indeed without the performance of any operations other than those of elementary algebra*.

---

* The following account of Hermite's method is taken in substance from Goursat's *Cours d'analyse mathématique* (first edition), t. 1, pp. 238–241.

Hermite's method depends upon a fundamental theorem in elementary algebra* which is also of great importance in the ordinary theory of partial fractions, viz. :

'*If X_1 and X_2 are two polynomials in x which have no common factor, and X_3 any third polynomial, then we can determine two polynomials A_1, A_2, such that*

$$A_1 X_1 + A_2 X_2 = X_3.'$$

Suppose that　　　　$Q(x) = Q_1 Q_2^2 Q_3^3 \ldots Q_t^t,$

$Q_1, \ldots$ denoting polynomials which have only simple roots and of which no two have any common factor. We can always determine $Q_1, \ldots$ by elementary methods, as is shown in the elements of the theory of equations†.

We can determine B and A_1 so that

$$B Q_1 + A_1 Q_2^2 Q_3^3 \ldots Q_t^t = P,$$

and therefore so that

$$R(x) = \frac{P}{Q} = \frac{A_1}{Q_1} + \frac{B}{Q_2^2 Q_3^3 \ldots Q_t^t}.$$

By a repetition of this process we can express $R(x)$ in the form

$$\frac{A_1}{Q_1} + \frac{A_2}{Q_2^2} + \ldots + \frac{A_t}{Q_t^t},$$

and the problem of the integration of $R(x)$ is reduced to that of the integration of a function

$$\frac{A}{Q^\nu},$$

where Q is a polynomial whose roots are all distinct. Since this is so, Q and its derived function Q' have no common factor : we can therefore determine C and D so that

$$CQ + DQ' = A.$$

Hence

$$\int \frac{A}{Q^\nu}\, dx = \int \frac{CQ + DQ'}{Q^\nu}\, dx$$

$$= \int \frac{C}{Q^{\nu-1}}\, dx - \frac{1}{\nu-1} \int D \frac{d}{dx}\left(\frac{1}{Q^{\nu-1}}\right) dx$$

$$= -\frac{D}{(\nu-1)\, Q^{\nu-1}} + \int \frac{E}{Q^{\nu-1}}\, dx,$$

where

$$E = C + \frac{D'}{\nu - 1}.$$

---

* See Chrystal's *Algebra*, vol. 1, pp. 119 *et seq.*

† See, for example, Hardy, *A course of pure mathematics* (2nd edition), p. 208.

Proceeding in this way, and reducing by unity at each step the power of $1/Q$ which figures under the sign of integration, we ultimately arrive at an equation

$$\int \frac{A}{Q^{\nu}}\,dx = R_{\nu}(x) + \int \frac{S}{Q}\,dx,$$

where R_{ν} is a rational function and S a polynomial.

The integral on the right-hand side has no rational part, since all the roots of Q are simple (§ 2). Thus the rational part of $\int R(x)\,dx$ is

$$R_{2}(x) + R_{3}(x) + \ldots + R_{t}(x),$$

and it has been determined without the need of any calculations other than those involved in the addition, multiplication and division of polynomials *.

5. (i) Let us consider, for example, the integral

$$\int \frac{4x^9 + 21x^6 + 2x^3 - 3x^2 - 3}{(x^7 - x + 1)^2}\,dx,$$

mentioned above (§ 3). We require polynomials A_1, A_2 such that

$$A_1 X_1 + A_2 X_2 = X_3 \quad \ldots\ldots\ldots\ldots\ldots\ldots\ldots\ldots(1),$$

where

$$X_1 = x^7 - x + 1, \quad X_2 = 7x^6 - 1, \quad X_3 = 4x^9 + 21x^6 + 2x^3 - 3x^2 - 3.$$

In general, if the degrees of X_1 and X_2 are m_1 and m_2, and that of X_3 does not exceed $m_1 + m_2 - 1$, we can suppose that the degrees of A_1 and A_2 do not exceed $m_2 - 1$ and $m_1 - 1$ respectively. For we know that polynomials B_1 and B_2 exist such that

$$B_1 X_1 + B_2 X_2 = X_3.$$

If B_1 is of degree not exceeding $m_2 - 1$, we take $A_1 = B_1$, and if it is of higher degree we write

$$B_1 = L_1 X_2 + A_1,$$

where A_1 is of degree not exceeding $m_2 - 1$. Similarly we write

$$B_2 = L_2 X_1 + A_2.$$

We have then

$$(L_1 + L_2) X_1 X_2 + A_1 X_1 + A_2 X_2 = X_3.$$

In this identity L_1 or L_2 or both may vanish identically, and in any case we see, by equating to zero the coefficients of the powers of x higher than the $(m_1 + m_2 - 1)$th, that $L_1 + L_2$ vanishes identically. Thus X_3 is expressed in the form required.

The actual determination of the coefficients in A_1 and A_2 is most easily performed by equating coefficients. We have then $m_1 + m_2$ linear equations

---

* The operation of forming the derived function of a given polynomial can of course be effected by a combination of these operations.

in the same number of unknowns. These equations must be consistent, since we know that a solution exists*.

If X_3 is of degree higher than $m_1 + m_2 - 1$, we must divide it by $X_1 X_2$ and express the remainder in the form required.

In this case we may suppose A_1 of degree 5 and A_2 of degree 6, and we find that
$$A_1 = -3x^2, \quad A_2 = x^3 + 3.$$
Thus the rational part of the integral is
$$-\frac{x^3 + 3}{x^7 - x + 1},$$
and, since $-3x^2 + (x^3 + 3)' = 0$, there is no transcendental part.

(ii) The following problem is instructive : *to find the conditions that*
$$\int \frac{ax^2 + 2\beta x + \gamma}{(Ax^2 + 2Bx + C)^2}\, dx$$
may be rational, and to determine the integral when it is rational.

We shall suppose that $Ax^2 + 2Bx + C$ is not a perfect square, as if it were the integral would certainly be rational. We can determine p, q and r so that
$$p(Ax^2 + 2Bx + C) + 2(qx + r)(Ax + B) = ax^2 + 2\beta x + \gamma,$$
and the integral becomes
$$p\int \frac{dx}{Ax^2 + 2Bx + C} - \int (qx + r)\frac{d}{dx}\left(\frac{1}{Ax^2 + 2Bx + C}\right) dx$$
$$= -\frac{qx + r}{Ax^2 + 2Bx + C} + (p + q)\int \frac{dx}{Ax^2 + 2Bx + C}.$$
The condition that the integral should be rational is therefore $p + q = 0$.

Equating coefficients we find
$$A(p + 2q) = a, \quad B(p + q) + Ar = \beta, \quad Cp + 2Br = \gamma.$$
Hence we deduce
$$p = -\frac{a}{A}, \qquad q = \frac{a}{A}, \qquad r = \frac{\beta}{A},$$
and $A\gamma + Ca = 2B\beta$. The condition required is therefore that the two quadratics $ax^2 + 2\beta x + \gamma$ and $Ax^2 + 2Bx + C$ should be harmonically related, and in this case
$$\int \frac{ax^2 + 2\beta x + \gamma}{(Ax^2 + 2Bx + C)^2}\, dx = -\frac{ax + \beta}{A(Ax^2 + 2Bx + C)}.$$

(iii) Another method of solution of this problem is as follows. If we write
$$Ax^2 + 2Bx + C = A(x - \lambda)(x - \mu),$$
and use the bilinear substitution
$$x = \frac{\lambda y + \mu}{y + 1},$$
then the integral is reduced to one of the form
$$\int \frac{ay^2 + 2by + c}{y^2}\, dy,$$

* It is easy to show that the solution is also unique.

and is rational if and only if $b=0$. But this is the condition that the quadratic $ay^2 + 2by + c$, corresponding to $ax^2 + 2\beta x + \gamma$, should be harmonically related to the degenerate quadratic y, corresponding to $Ax^2 + 2Bx + C$. The result now follows from the fact that harmonic relations are not changed by bilinear transformation.

It is not difficult to show, by an adaptation of this method, that

$$\int \frac{(ax^2 + 2\beta x + \gamma)(a_1 x^2 + 2\beta_1 x + \gamma_1) \dots (a_n x^2 + 2\beta_n x + \gamma_n)}{(Ax^2 + 2Bx + C)^{n+2}}\, dx$$

is rational if all the quadratics are harmonically related to any one of those in the numerator. This condition is sufficient but not necessary.

(iv) As a further example of the use of the method (ii) the reader may show that *the necessary and sufficient condition that*

$$\int \frac{f(x)}{\{F(x)\}^2}\, dx,$$

where f and F are polynomials with no common factor, and F has no repeated factor, should be rational, is that $f'F' - fF''$ should be divisible by F.

6. It appears from the preceding paragraphs that we can always find the rational part of the integral, and can find the complete integral if we can find the roots of $Q(x) = 0$. The question is naturally suggested as to the maximum of information which can be obtained about the logarithmic part of the integral in the general case in which the factors of the denominator cannot be determined explicitly. For there are polynomials which, although they cannot be completely resolved into such factors, can nevertheless be partially resolved. For example

$$x^{14} - 2x^8 - 2x^7 - x^4 - 2x^3 + 2x + 1 = (x^7 + x^2 - 1)(x^7 - x^2 - 2x - 1),$$

$$x^{14} - 2x^8 - 2x^7 - 2x^4 - 4x^3 - x^2 + 2x + 1$$
$$= \{x^7 + x^2 \sqrt{2} + x(\sqrt{2} - 1) - 1\}\{x^7 - x^2 \sqrt{2} - x(\sqrt{2} + 1) - 1\}.$$

The factors of the first polynomial have rational coefficients : in the language of the theory of equations, the polynomial is *reducible in the rational domain*. The second polynomial is reducible in the domain formed by the *adjunction* of the single irrational $\sqrt{2}$ to the rational domain*.

We may suppose that every possible decomposition of $Q(x)$ of this nature has been made, so that

$$Q = Q_1 Q_2 \dots Q_t.$$

---

* See Cajori, *An introduction to the modern theory of equations* (Macmillan, 1904) ; Mathews, *Algebraic equations* (*Cambridge tracts in mathematics*, no. 6), pp. 6–7.

Then we can resolve $R(x)$ into a sum of partial fractions of the type

$$\int \frac{P_\nu}{Q_\nu}\,dx,$$

and so we need only consider integrals of the type

$$\int \frac{P}{Q}\,dx,$$

where no further resolution of Q is possible or, in technical language, *Q is irreducible by the adjunction of any algebraical irrationality.*

Suppose that this integral can be evaluated in a form involving only constants which can be expressed explicitly in terms of the constants which occur in P/Q. It must be of the form

$$A_1 \log X_1 + \ldots + A_k \log X_k \ldots\ldots\ldots\ldots\ldots(1),$$

where the A's are constants and the X's polynomials. We can suppose that no X has any repeated factor ξ^m, where ξ is a polynomial. For such a factor could be determined rationally in terms of the co-efficients of X, and the expression (1) could then be modified by taking out the factor ξ^m from X and inserting a new term $mA \log \xi$. And for similar reasons we can suppose that no two X's have any factor in common.

Now $$\frac{P}{Q} = A_1 \frac{X_1'}{X_1} + A_2 \frac{X_2'}{X_2} + \ldots + A_k \frac{X_k'}{X_k},$$

or $$PX_1 X_2 \ldots X_k = Q \Sigma A_\nu X_1 \ldots X_{\nu-1} X_\nu' X_{\nu+1} \ldots X_k.$$

All the terms under the sign of summation are divisible by X_1 save the first, which is prime to X_1. Hence Q must be divisible by X_1: and similarly, of course, by $X_2, X_3, \ldots, X_k$. But, since P is prime to Q, $X_1 X_2 \ldots X_k$ is divisible by Q. Thus Q must be a constant multiple of $X_1 X_2 \ldots X_k$. But Q is *ex hypothesi* not resoluble into factors which contain only explicit algebraical irrationalities. Hence all the X's save one must reduce to constants, and so P must be a constant multiple of Q', and

$$\int \frac{P}{Q}\,dx = A \log Q,$$

where A is a constant. Unless this is the case the integral cannot be expressed in a form involving only constants expressed explicitly in terms of the constants which occur in P and Q.

Thus, for instance, the integral

$$\int \frac{dx}{x^5 + ax + b}$$

cannot, except in special cases*, be expressed in a form involving only constants expressed explicitly in terms of a and b; and the integral

$$\int \frac{5x^4+c}{x^5+ax+b}\,dx$$

can in general be so expressed if and only if $c=a$. We thus confirm an inference made before (§ 3) in a less accurate way.

Before quitting this part of our subject we may consider one further problem : *under what circumstances is*

$$\int R(x)\,dx = A \log R_1(x)$$

where A is a constant and R_1 rational? Since the integral has no rational part, it is clear that $Q(x)$ must have only simple factors, and that the degree of $P(x)$ must be less than that of $Q(x)$. We may therefore use the formula

$$\int R(x)\,dx = \log \prod_1^r \{(x-a_s)^{P(a_s)/Q'(a_s)}\}.$$

The necessary and sufficient condition is that all the numbers $P(a_s)/Q'(a_s)$ should be commensurable. If *e.g.*

$$R(x) = \frac{x-\gamma}{(x-a)(x-\beta)},$$

then $(a-\gamma)/(a-\beta)$ and $(\beta-\gamma)/(\beta-a)$ must be commensurable, *i.e.* $(a-\gamma)/(\beta-\gamma)$ must be a rational number. If the denominator is given we can find all the values of γ which are admissible : for $\gamma=(aq-\beta p)/(q-p)$, where p and q are integers.

7. Our discussion of the integration of rational functions is now complete. It has been throughout of a theoretical character. We have not attempted to consider what are the simplest and quickest methods for the actual calculation of the types of integral which occur most commonly in practice. This problem lies outside our present range: the reader may consult

O. Stolz, *Grundzüge der Differential-und-integralrechnung*, vol. 1, ch. 7 :

J. Tannery, *Leçons d'algèbre et d'analyse*, vol. 2, ch. 18 :

Ch.-J. de la Vallée-Poussin, *Cours d'analyse*, ed. 3, vol. 1, ch. 5 :

T. J. I'A. Bromwich, *Elementary integrals* (Bowes and Bowes, 1911):

G. H. Hardy, *A course of pure mathematics*, ed. 2, ch. 6.

* The equation $x^5+ax+b=0$ is soluble by radicals in certain cases. See Mathews, *l.c.*, pp. 52 *et seq.*

V. Algebraical Functions

1. We shall now consider the integrals of algebraical functions, explicit or implicit.   The theory of the integration of such functions is far more extensive and difficult than that of rational functions, and we can give here only a brief account of a few of the most important results and of the most obvious of their applications.

If $y_1, y_2, \ldots, y_n$ are algebraical functions of x, then any algebraical function z of $x, y_1, \ldots, y_n$ is an algebraical function of x.   This is obvious if we confine ourselves to *explicit* algebraical functions.   In the general case we have a number of equations of the type

$$P_{\nu,0}(x)\, y_\nu^{m_\nu} + P_{\nu,1}(x)\, y_\nu^{m_\nu-1} + \ldots + P_{\nu,m_\nu}(x) = 0 \quad (\nu = 1, 2, \ldots, n),$$

and $\qquad P_0(x, y_1, \ldots, y_n)\, z^m + \ldots + P_m(x, y_1, \ldots, y_n) = 0,$

where the P's represent polynomials in their arguments.   The elimination of $y_1, y_2, \ldots, y_n$ between these equations gives an equation in z whose coefficients are polynomials in x only.

The importance of this from our present point of view lies in the fact that we may consider the standard algebraical integral under any of the forms

$$\int y\, dx,$$

where $f(x, y) = 0$;

$$\int R(x, y)\, dx,$$

where $f(x, y) = 0$ and R is rational ; or

$$\int R(x, y_1, \ldots, y_n)\, dx,$$

where $f_1(x, y) = 0, \ldots, f_n(x, y_n) = 0$.   It is, for example, much more convenient to treat such an irrational as

$$\frac{x - \sqrt{(x+1)} - \sqrt{(x-1)}}{1 + \sqrt{(x+1)} + \sqrt{(x-1)}}$$

as a rational function of x, y_1, y_2, where $y_1 = \sqrt{(x+1)}$, $y_2 = \sqrt{(x-1)}$, $y_1^2 = x+1$, $y_2^2 = x-1$, than as a rational function of x and y, where

$$y = \sqrt{(x+1)} + \sqrt{(x-1)},$$
$$y^4 - 4xy^2 + 4 = 0.$$

To treat it as a simple irrational y, so that our fundamental equation is

$$(x-y)^4 - 4x(x-y)^2(1+y)^2 + 4(1+y)^4 = 0$$

is evidently the least convenient course of all.

Before we proceed to consider the general form of the integral of an algebraical function we shall consider one most important case in which the integral can be at once reduced to that of a rational function, and is therefore always an elementary function itself.

2. The class of integrals alluded to immediately above is that covered by the following theorem.

If there is a variable t connected with x and y (or $y_1, y_2, \ldots, y_n$) by rational relations
$$x = R_1(t), \quad y = R_2(t)$$
(or $y_1 = R_2^{(1)}(t), y_2 = R_2^{(2)}(t), \ldots$), then the integral
$$\int R(x, y)\, dx$$
(or $\int R(x, y_1, \ldots, y_n)\, dx$) is an elementary function.

The truth of this proposition follows immediately from the equations
$$R(x, y) = R\{R_1(t), R_2(t)\} = S(t),$$
$$\frac{dx}{dt} = R_1'(t) = T(t),$$
$$\int R(x, y)\, dx = \int S(t)\, T(t)\, dt = \int U(t)\, dt,$$
where all the capital letters denote rational functions.

The most important case of this theorem is that in which x and y are connected by the general quadratic relation
$$(a, b, c, f, g, h \Ydown x, y, 1)^2 = 0.$$
The integral can then be made rational in an infinite number of ways For suppose that (ξ, η) is any point on the conic, and that
$$(y - \eta) = t(x - \xi)$$
is any line through the point. If we eliminate y between these equations, we obtain an equation of the second degree in x, say
$$T_0 x^2 + 2T_1 x + T_2 = 0,$$
where T_0, T_1, T_2 are polynomials in t. But one root of this equation must be ξ, which is independent of t; and when we divide by $x - \xi$ we obtain an equation of the *first* degree for the abscissa of the variable point of intersection, in which the coefficients are again polynomials in t. Hence this abscissa is a rational function of t; the ordinate of the point is also a rational function of t, and as t varies this point

coincides with every point of the conic in turn. In fact the equation of the conic may be written in the form

$$au^2 + 2huv + bv^2 + 2\left(a\xi + h\eta + g\right)u + 2\left(h\xi + b\eta + f\right)v = 0,$$

where $u = x - \xi$, $v = y - \eta$, and the other point of intersection of the line $v = tu$ and the conic is given by

$$x = \xi - \frac{2\left\{a\xi + h\eta + g + t\left(h\xi + b\eta + f\right)\right\}}{a + 2ht + bt^2},$$

$$y = \eta - \frac{2t\left\{a\xi + h\eta + g + t\left(h\xi + b\eta + f\right)\right\}}{a + 2ht + bt^2}.$$

An alternative method is to write

$$ax^2 + 2hxy + by^2 = b\left(y - \mu x\right)\left(y - \mu' x\right),$$

so that $y - \mu x = 0$ and $y - \mu' x = 0$ are parallel to the asymptotes of the conic, and to put

$$y - \mu x = t.$$

Then
$$y - \mu' x = -\frac{2gx + 2fy + c}{bt};$$

and from these two equations we can calculate x and y as rational functions of t. The principle of this method is of course the same as that of the former method : (ξ, η) is now at infinity, and the pencil of lines through (ξ, η) is replaced by a pencil parallel to an asymptote.

The most important case is that in which $b = -1, f = h = 0$, so that

$$y^2 = ax^2 + 2gx + c.$$

The integral is then made rational by the substitution

$$x = \xi - \frac{2\left(a\xi + g - t\eta\right)}{a - t^2}, \quad y = \eta - \frac{2t\left(a\xi + g - t\eta\right)}{a - t^2}$$

where ξ, η are any numbers such that

$$\eta^2 = a\xi^2 + 2g\xi + c.$$

We may for instance suppose that $\xi = 0$, $\eta = \sqrt{c}$; or that $\eta = 0$, while ξ is a root of the equation $a\xi^2 + 2g\xi + c = 0$. Or again the integral is made rational by putting $y - x\sqrt{a} = t$, when

$$x = -\frac{t^2 - c}{2\left(t\sqrt{a} - g\right)}, \quad y = \frac{\left(t^2 + c\right)\sqrt{a} - 2gt}{2\left(t\sqrt{a} - g\right)}.$$

3. We shall now consider in more detail the problem of the calculation of

$$\int R\left(x, y\right)dx,$$

where
$$y = \sqrt{X} = \sqrt{\left(ax^2 + 2bx + c\right)}*$$

---

* We now write b for g for the sake of symmetry in notation.

The most interesting case is that in which a, b, c and the constants which occur in R are real, and we shall confine our attention to this case.

Let
$$R(x, y) = \frac{P(x, y)}{Q(x, y)},$$

where P and Q are polynomials. Then, by means of the equation

$$y^2 = ax^2 + 2bx + c,$$

$R(x, y)$ may be reduced to the form

$$\frac{A + B\sqrt{X}}{C + D\sqrt{X}} = \frac{(A + B\sqrt{X})(C - D\sqrt{X})}{C^2 - D^2 X},$$

where A, B, C, D are polynomials in x; and so to the form $M + N\sqrt{X}$, where M and N are rational, or (what is the same thing) the form

$$P + \frac{Q}{\sqrt{X}},$$

where P and Q are rational. The rational part may be integrated by the methods of section IV., and the integral

$$\int \frac{Q}{\sqrt{X}}\, dx$$

may be reduced to the sum of a number of integrals of the forms

$$\int \frac{x^r}{\sqrt{X}}\, dx, \qquad \int \frac{dx}{(x - p)^r \sqrt{X}}, \qquad \int \frac{\xi x + \eta}{(ax^2 + 2\beta x + \gamma)^r \sqrt{X}}\, dx \quad \ldots\ldots(1),$$

where p, ξ, η, a, β, γ are real constants and r a positive integer. The result is generally required in an explicitly real form : and, as further progress depends on transformations involving p (or a, β, γ), it is generally not advisable to break up a quadratic factor $ax^2 + 2\beta x + \gamma$ into its constituent linear factors when these factors are complex.

All of the integrals (1) may be reduced, by means of elementary formulae of reduction*, to dependence upon three fundamental integrals, viz.

$$\int \frac{dx}{\sqrt{X}}, \qquad \int \frac{dx}{(x - p)\sqrt{X}}, \qquad \int \frac{\xi x + \eta}{(ax^2 + 2\beta x + \gamma)\sqrt{X}}\, dx \quad \ldots\ldots\ldots(2).$$

4. The first of these integrals may be reduced, by a substitution of the type $x = t + k$, to one or other of the three standard forms

$$\int \frac{dt}{\sqrt{(m^2 - t^2)}}, \qquad \int \frac{dt}{\sqrt{(t^2 + m^2)}}, \qquad \int \frac{dt}{\sqrt{(t^2 - m^2)}},$$

where $m > 0$. These integrals may be rationalised by the substitutions

$$t = \frac{2mu}{1 + u^2}, \qquad t = \frac{2mu}{1 - u^2}, \qquad t = \frac{m(1 + u^2)}{2u};$$

but it is simpler to use the transcendental substitutions

$$t = m \sin \phi, \qquad t = m \sinh \phi, \qquad t = m \cosh \phi.$$

---

* See, for example, Bromwich, *l.c.*, pp. 16 *et seq.*

These last substitutions are generally the most convenient for the reduction of an integral which contains one or other of the irrationalities

$$\sqrt{(m^2-t^2)}, \qquad \sqrt{(t^2+m^2)}, \qquad \sqrt{(t^2-m^2)},$$

though the alternative substitutions

$$t=m\tanh\phi, \qquad t=m\tan\phi, \qquad t=m\sec\phi$$

are often useful.

It has been pointed out by Dr Bromwich that the forms usually given in text-books for these three standard integrals, viz.

$$\text{arc sin}\frac{t}{m}, \qquad \text{arg sinh}\frac{t}{m} \qquad \text{arg cosh}\frac{t}{m}.$$

are not quite accurate. It is obvious, for example, that the first two of these functions are odd functions of m, while the corresponding integrals are even functions. The correct formulae are

$$\text{arc sin}\frac{t}{|m|}, \qquad \text{arg sinh}\frac{t}{|m|}=\log\frac{t+\sqrt{(t^2+m^2)}}{|m|}$$

and

$$\pm\,\text{arg cosh}\frac{|t|}{|m|}=\log\left|\frac{t+\sqrt{(t^2-m^2)}}{m}\right|,$$

where the ambiguous sign is the same as that of t. It is in some ways more convenient to use the equivalent forms

$$\text{arc tan}\frac{t}{\sqrt{(m^2-t^2)}}, \qquad \text{arg tanh}\frac{t}{\sqrt{(t^2+m^2)}}, \qquad \text{arg tanh}\frac{t}{\sqrt{(t^2-m^2)}}.$$

5. The integral
$$\int\frac{dx}{(x-p)\sqrt{X}}$$

may be evaluated in a variety of ways.

If p is a root of the equation $X=0$, then X may be written in the form $a(x-p)(x-q)$, and the value of the integral is given by one or other of the formulae

$$\int\frac{dx}{(x-p)\sqrt{\{(x-p)(x-q)\}}}=\frac{2}{q-p}\sqrt{\left(\frac{x-q}{x-p}\right)},$$

$$\int\frac{dx}{(x-p)^{5/2}}=-\frac{2}{3(x-p)^{3/2}}.$$

We may therefore suppose that p is not a root of $X=0$.

(i) We may follow the general method described above, taking

$$\xi=p, \qquad \eta=\sqrt{(ap^2+2bp+c)}*.$$

Eliminating y from the equations

$$y^2=ax^2+2bx+c, \qquad y-\eta=t(x-\xi),$$

and dividing by $x-\xi$, we obtain

$$t^2(x-\xi)+2\eta t-a(x+\xi)-2b=0,$$

and so

$$-\frac{2dt}{t^2-a}=\frac{dx}{t(x-\xi)+\eta}=\frac{dx}{y}.$$

* Cf. Jordan, *Cours d'analyse*, ed. 2, vol. 2, p. 21.

Hence
$$\int \frac{dx}{(x-\xi)\,y} = -2\int \frac{dt}{(x-\xi)\,(t^2-a)}.$$

But
$$(t^2-a)\,(x-\xi) = 2a\xi + 2b - 2\eta t\ ;$$

and so
$$\int \frac{dx}{(x-p)\,y} = -\int \frac{dt}{a\xi+b-\eta t} = \frac{1}{\eta}\log(a\xi+b-\eta t)$$
$$= \frac{1}{\sqrt{(ap^2+2bp+c)}}\log\{t\sqrt{(ap^2+2bp+c)}-ap-b\}.$$

If $ap^2+2bp+c<0$ the transformation is imaginary.

Suppose, e.g., (a) $y=\sqrt{(x+1)}$, $p=0$, or (b) $y=\sqrt{(x-1)}$, $p=0$. We find

(a)
$$\int \frac{dx}{x\sqrt{(x+1)}} = \log(t-\tfrac{1}{2}),$$

where
$$t^2x+2t-1=0,$$

or
$$t=\frac{-1+\sqrt{(x+1)}}{x}\ ;$$

and

(b)
$$\int \frac{dx}{x\sqrt{(x-1)}} = -i\log(it-\tfrac{1}{2}),$$

where
$$t^2x+2it-1=0.$$

Neither of these results is expressed in the simplest form, the second in particular being very inconvenient.

(ii) The most straightforward method of procedure is to use the substitution
$$x-p=\frac{1}{t}.$$

We then obtain
$$\int \frac{dx}{(x-p)\,y} = \int \frac{dt}{\sqrt{(a_1 t^2+2b_1 t+c_1)}}\ ,$$

where a_1, b_1, c_1 are certain simple functions of a, b, c, and p. The further reduction of this integral has been discussed already.

(iii) A third method of integration is that adopted by Sir G. Greenhill[*], who uses the transformation
$$t=\frac{\sqrt{(ax^2+2bx+c)}}{x-p}.$$

It will be found that
$$\int \frac{dx}{(x-p)\sqrt{X}} = \int \frac{dt}{\sqrt{\{(ap^2+2bp+c)\,t^2+b^2-ac\}}}\ ,$$

which is of one of the three standard forms mentioned in § 4.

[*] A. G. Greenhill, *A chapter in the integral calculus* (Francis Hodgson, 1888), p. 12 : *Differential and integral calculus*, p. 399.

6. It remains to consider the integral

$$\int \frac{\xi x + \eta}{(ax^2 + 2\beta x + \gamma)\sqrt{X}}\, dx = \int \frac{\xi x + \eta}{X_1 \sqrt{X}}\, dx,$$

where $ax^2 + 2\beta x + \gamma$ or X_1 is a quadratic with complex linear factors. Here again there is a choice of methods at our disposal.

We may suppose that X_1 is not a constant multiple of X. If it is, then the value of the integral is given by the formula

$$\int \frac{\xi x + \eta}{(ax^2 + 2bx + c)^{3/2}}\, dx = \frac{\eta(ax+b) - \xi(bx+c)}{\sqrt{\{(ac - b^2)(ax^2 + 2bx + c)\}}} \quad *.$$

(i) The standard method is to use the substitution

$$x = \frac{\mu t + \nu}{t + 1} \quad \dots\dots\dots\dots\dots\dots\dots\dots\dots(1),$$

where μ and ν are so chosen that

$$a\mu\nu + b(\mu+\nu) + c = 0, \qquad a\mu\nu + \beta(\mu+\nu) + \gamma = 0 \quad \dots\dots\dots(2).$$

The values of μ and ν which satisfy these conditions are the roots of the quadratic

$$(a\beta - ba)\mu^2 - (ca - a\gamma)\mu + (b\gamma - c\beta) = 0.$$

The roots will be real and distinct if

$$(ca - a\gamma)^2 > 4(a\beta - ba)(b\gamma - c\beta),$$

or if $\qquad (a\gamma + ca - 2b\beta)^2 > 4(ac - b^2)(a\gamma - \beta^2) \quad \dots\dots\dots\dots(3).$

Now $a\gamma - \beta^2 > 0$, so that (3) is certainly satisfied if $ac - b^2 < 0$. But if $ac - b^2$ and $a\gamma - \beta^2$ are both positive then $a\gamma$ and ca have the same sign, and

$$(a\gamma + ca - 2b\beta)^2 \geq (|a\gamma + ca| - 2|b\beta|)^2 > 4\{\sqrt{(aca\gamma)} - |b\beta|\}^2$$
$$= 4[(ac - b^2)(a\gamma - \beta^2) + \{|b|\sqrt{(a\gamma)} - |\beta|\sqrt{(ac)}\}^2]$$
$$\geq 4(ac - b^2)(a\gamma - \beta^2).$$

Thus the values of μ and ν are in any case real and distinct.

It will be found, on carrying out the substitution (1), that

$$\int \frac{\xi x + \eta}{X_1 \sqrt{X}}\, dx = H \int \frac{t\, dt}{(\mathbf{A}t^2 + \mathbf{B})\sqrt{(At^2 + B)}} + K \int \frac{dt}{(\mathbf{A}t^2 + \mathbf{B})\sqrt{(At^2 + B)}},$$

where $\mathbf{A}, \mathbf{B}, A, B, H,$ and K are constants. Of these two integrals, the first is rationalised by the substitution

$$\frac{1}{\sqrt{(At^2 + B)}} = u,$$

and the second by the substitution

$$\frac{t}{\sqrt{(At^2 + B)}} = v. \dagger$$

It should be observed that this method fails in the special case in which

---

* Bromwich, *l.c.*, p. 16.

† The method sketched here is that followed by Stolz (see the references given on p. 21). Dr Bromwich's method is different in detail but the same in principle.

$a\beta - ba = 0$. In this case, however, the substitution $ax + b = t$ reduces the integral to one of the form

$$\int \frac{Ht + K}{(\mathbf{A}t^2 + \mathbf{B})\sqrt{(At^2 + B)}}\, dt,$$

and the reduction may then be completed as before.

(ii)   An alternative method is to use Sir G. Greenhill's substitution

$$t = \sqrt{\left(\frac{X}{ax^2 + 2\beta x + \gamma}\right)} = \sqrt{\left(\frac{X}{X_1}\right)}.$$

If

$$J = (a\beta - ba)\, x^2 - (ca - a\gamma)\, x + (b\gamma - c\beta),$$

then

$$\frac{1}{t}\frac{dt}{dx} = \frac{J}{XX_1} \quad \dots\dots\dots\dots\dots\dots\dots\dots\dots(1).$$

The maximum and minimum values of t are given by $J = 0$.

Again

$$t^2 - \lambda = \frac{(a - \lambda a)\, x^2 + 2\,(b - \lambda\beta)\, x + (c - \lambda\gamma)}{X}\;;$$

and the numerator will be a perfect square if

$$K = (a\gamma - \beta^2)\,\lambda^2 - (a\gamma + ca - 2b\beta)\,\lambda + (ac - b^2) = 0.$$

It will be found by a little calculation that the discriminant of this quadratic and that of J differ from one another and from

$$(\phi - \phi_1)\,(\phi - \phi_1')\,(\phi' - \phi_1)\,(\phi' - \phi_1'),$$

where ϕ, ϕ' are the roots of $X = 0$ and ϕ_1, ϕ_1' those of $X_1 = 0$, only by a constant factor which is always negative. Since ϕ_1 and ϕ_1' are conjugate complex numbers, this product is positive, and so $J = 0$ and $K = 0$ have real roots*. We denote the roots of the latter by

$$\lambda_1, \lambda_2 \quad (\lambda_1 > \lambda_2).$$

Then

$$\lambda_1 - t^2 = \frac{\{x\sqrt{(\lambda_1 a - a)} + \sqrt{(\lambda_1\gamma - c)}\}^2}{X_1} = \frac{(mx + n)^2}{X_1} \quad \dots\dots\dots(2),$$

$$t^2 - \lambda_2 = \frac{\{x\sqrt{(a - \lambda_2 a)} + \sqrt{(c - \lambda_2\gamma)}\}^2}{X_1} = \frac{(m'x + n')^2}{X_1} \quad \dots\dots\dots(2'),$$

say. Further, since $t^2 - \lambda$ can vanish for two equal values of x only if λ is equal to λ_1 or λ_2, i.e. when t is a maximum or a minimum, J can differ from

$$(mx + n)\,(m'x + n')$$

only by a constant factor; and by comparing coefficients and using the identity

$$(\lambda_1 a - a)\,(a - \lambda_2 a) = \frac{(a\beta - ba)^2}{a\gamma - \beta^2},$$

we find that

$$J = \sqrt{(a\gamma - \beta^2)}\,(mx + n)\,(m'x + n') \quad \dots\dots\dots\dots\dots(3).$$

Finally, we can write $\xi x + \eta$ in the form

$$A\,(mx + n) + B\,(m'x + n').$$

---

* That the roots of $J = 0$ are real has been proved already (p. 28) in a different manner.

Using equations (1), (2), (2′), and (3), we find that

$$\int \frac{\xi x + \eta}{X_1 \sqrt{X}}\, dx = \int \frac{A\,(mx+n)+B\,(m'x+n')}{J}\, \sqrt{X_1}\, dt$$

$$= \frac{A}{\sqrt{(\alpha\gamma-\beta^2)}} \int \frac{dt}{\sqrt{(\lambda_1 - t^2)}} + \frac{B}{\sqrt{(\alpha\gamma-\beta^2)}} \int \frac{dt}{\sqrt{(t^2-\lambda_2)}},$$

and the integral is reduced to a sum of two standard forms.

This method is very elegant, and has the advantage that the whole work of transformation is performed in one step. On the other hand it is somewhat artificial, and it is open to the logical objection that it introduces the root $\sqrt{X_1}$, which, in virtue of Laplace's principle (III., 2), cannot really be involved in the final result*.

7. We may now proceed to consider the general case to which the theorem of IV., § 2 applies. It will be convenient to recall two well-known definitions in the theory of algebraical plane curves. A curve of degree n can have at most $\frac{1}{2}(n-1)(n-2)$ double points†. If the actual number of double points is ν, then the number

$$p = \tfrac{1}{2}(n-1)(n-2) - \nu$$

is called the *deficiency*‡ of the curve.

If the coordinates x, y of the points on a curve can be expressed *rationally* in terms of a parameter t by means of equations

$$x = R_1(t), \quad y = R_2(t),$$

then we shall say that the curve is *unicursal*. In this case we have seen that we can always evaluate

$$\int R(x, y)\, dx$$

in terms of elementary functions.

The fundamental theorem in this part of our subject is

'*A curve whose deficiency is zero is unicursal, and vice versa*'.

Suppose first that the curve possesses the maximum number of double points§. Since

$$\tfrac{1}{2}(n-1)(n-2)+n-3 = \tfrac{1}{2}(n-2)(n+1)-1,$$

---

* The superfluous root may be eliminated from the result by a trivial transformation, just as $\sqrt{(1+x^2)}$ may be eliminated from

$$\text{arc sin } \frac{x}{\sqrt{(1+x^2)}}$$

by writing this function in the form arc tan x.

† Salmon, *Higher plane curves*, p. 29.

‡ Salmon, *ibid.*, p. 29. French *genre*, German *Geschlecht*.

§ We suppose in what follows that the singularities of the curve are all ordinary nodes. The necessary modifications when this is not the case are not difficult to

and $\frac{1}{2}(n-2)(n+1)$ points are just sufficient to determine a curve of degree $n-2$*, we can draw, through the $\frac{1}{2}(n-1)(n-2)$ double points and $n-3$ other points chosen arbitrarily on the curve, a simply infinite set of curves of degree $n-2$, which we may suppose to have the equation

$$g(x, y) + t\,h(x, y) = 0,$$

where t is a variable parameter and $g = 0$, $h = 0$ are the equations of two particular members of the set. Any one of these curves meets the given curve in $n(n-2)$ points, of which $(n-1)(n-2)$ are accounted for by the $\frac{1}{2}(n-1)(n-2)$ double points, and $n-3$ by the other $n-3$ arbitrarily chosen points. These

$$(n-1)(n-2) + n - 3 = n(n-2) - 1$$

points are independent of t; and so there is but *one* point of intersection which depends on t. The coordinates of this point are given by

$$g(x, y) + t\,h(x, y) = 0, \qquad f(x, y) = 0.$$

The elimination of y gives an equation of degree $n(n-2)$ in x, whose coefficients are polynomials in t; and but one root of this equation varies with t. The eliminant is therefore divisible by a factor of degree $n(n-2) - 1$ which does not contain t. There remains a simple equation in x whose coefficients are polynomials in t. Thus the x-coordinate of the variable point is determined as a rational function of t, and the y-coordinate may be similarly determined.

We may therefore write

$$x = R_1(t), \qquad y = R_2(t).$$

If we reduce these fractions to the same denominator, we express the coordinates in the form

$$x = \frac{\phi_1(t)}{\phi_3(t)}, \qquad y = \frac{\phi_2(t)}{\phi_3(t)} \quad \dots\dots\dots\dots\dots(1),$$

where ϕ_1, ϕ_2, ϕ_3 are polynomials which have no common factor. The polynomials will in general be of degree n; none of them can be of

make. An ordinary multiple point of order k may be regarded as equivalent to $\frac{1}{2}k(k-1)$ ordinary double points. A curve of degree n which has an ordinary multiple point of order $n-1$, equivalent to $\frac{1}{2}(n-1)(n-2)$ ordinary double points, is therefore unicursal. The theory of higher plane curves abounds in puzzling particular cases which have to be fitted into the general theory by more or less obvious conventions, and to give a satisfactory account of a complicated compound singularity is sometimes by no means easy. In the investigation which follows we confine ourselves to the simplest case.

* Salmon, *l.c.*, p. 16.

higher degree, and one at least must be actually of that degree, since an arbitrary straight line

$$\lambda x + \mu y + \nu = 0$$

must cut the curve in exactly n points*.

We can now prove the second part of the theorem.   If

$$x : y : 1 :: \phi_1(t) : \phi_2(t) : \phi_3(t),$$

where ϕ_1, ϕ_2, ϕ_3 are polynomials of degree n, then the line

$$ux + vy + w = 0$$

will meet the curve in n points whose parameters are given by

$$u\phi_1(t) + v\phi_2(t) + w\phi_3(t) = 0.$$

This equation will have a double root t_0 if

$$u\phi_1(t_0) + v\phi_2(t_0) + w\phi_3(t_0) = 0,$$
$$u\phi_1'(t_0) + v\phi_2'(t_0) + w\phi_3'(t_0) = 0.$$

Hence the equation of the tangent at the point t_0 is

$$\begin{vmatrix} x & y & 1 \\ \phi_1(t_0) & \phi_2(t_0) & \phi_3(t_0) \\ \phi_1'(t_0) & \phi_2'(t_0) & \phi_3'(t_0) \end{vmatrix} = 0 \quad \ldots\ldots\ldots\ldots(2).$$

If (x, y) is a fixed point, then the equation (2) may be regarded as an equation to determine the parameters of the points of contact of the tangents from (x, y).   Now

$$\phi_2(t_0)\,\phi_3'(t_0) - \phi_2'(t_0)\,\phi_3(t_0)$$

is of degree $2n - 2$ in t_0, the coefficient of t_0^{2n-1} obviously vanishing. Hence in general the number of tangents which can be drawn to a unicursal curve from a fixed point (the *class* of the curve) is $2n - 2$. But the class of a curve whose only singular points are δ nodes is known† to be $n(n-1) - 2\delta$.   Hence the number of nodes is

$$\tfrac{1}{2}\{n(n-1) - (2n-2)\} = \tfrac{1}{2}(n-1)(n-2).$$

It is perhaps worth pointing out how the proof which precedes requires modification if some only of the singular points are nodes and the rest ordinary cusps. The first part of the proof remains unaltered. The equation

---

* See Niewenglowski's *Cours de géométrie analytique*, vol. 2, p. 103. By way of illustration of the remark concerning particular cases in the footnote (§) to page 30, the reader may consider the example given by Niewenglowski in which

$$x = \frac{t^2}{t^2-1}, \quad y = \frac{t^2+1}{t^2-1};$$

equations which appear to represent the straight line $2x = y + 1$ (part of the line only, if we consider only real values of t).

† Salmon, *l.c.*, p. 54.

(2) must now be regarded as giving the values of t which correspond to (a) points at which the tangent passes through (x, y) and (b) cusps, since any line through a cusp 'cuts the curve in two coincident points'*. We have therefore

$$2n - 2 = m + \kappa,$$

where m is the class of the curve. But

$$m = n(n-1) - 2\delta - 3\kappa, \dagger$$

and so

$$\delta + \kappa = \tfrac{1}{2}(n-1)(n-2). \ddagger$$

8. (i) The preceding argument fails if $n < 3$, but we have already seen that all conics are unicursal. The case next in importance is that of a cubic with a double point. If the double point is not at infinity we can, by a change of origin, reduce the equation of the curve to the form

$$(ax + by)(cx + dy) = px^3 + 3qx^2y + 3rxy^2 + sy^3 ;$$

and, by considering the intersections of the curve with the line $y = tx$, we find

$$x = \frac{(a + bt)(c + dt)}{p + 3qt + 3rt^2 + st^3}, \qquad y = \frac{t(a + bt)(c + dt)}{p + 3qt + 3rt^2 + st^3}.$$

If the double point is at infinity, the equation of the curve is of the form

$$(\alpha x + \beta y)^2(\gamma x + \delta y) + \epsilon x + \zeta y + \theta = 0,$$

the curve having a pair of parallel asymptotes; and, by considering the intersection of the curve with the line $\alpha x + \beta y = t$, we find

$$x = -\frac{\delta t^3 + \zeta t + \beta\theta}{(\beta\gamma - \alpha\delta)t^2 + \epsilon\beta - \alpha\zeta}, \qquad y = \frac{\gamma t^3 + \epsilon t + \alpha\theta}{(\beta\gamma - \alpha\delta)t^3 + \epsilon\beta - \alpha\zeta}.$$

(ii) The case next in complexity is that of a quartic with three double points.

(a) The lemniscate $(x^2 + y^2)^2 = a^2(x^2 - y^2)$

has three double points, the origin and the circular points at infinity. The circle

$$x^2 + y^2 = t(x - y)$$

---

* This means of course that the equation obtained by substituting for x and y, in the equation of the line, their parametric expressions in terms of t, has a repeated root. This property is possessed by the tangent at an ordinary point and by any line through a cusp, but not by any line through a node except the two tangents.

† Salmon, *l.c.*, p. 65.

‡ I owe this remark to Mr A. B. Mayne. Dr Bromwich has however pointed out to me that substantially the same argument is given by Mr W. A. Houston, 'Note on unicursal plane curves', *Messenger of mathematics*, vol. 28, 1899, pp. 187–189.

passes through these points and one other fixed point at the origin, as it touches the curve there. Solving, we find

$$x = \frac{a^2 t\,(t^2 + a^2)}{t^4 + a^4}, \qquad y = \frac{a^2 t\,(t^2 - a^2)}{t^4 + a^4}.$$

(b) The curve $2ay^3 - 3a^2 y^2 = x^4 - 2a^2 x^2$

has the double points $(0, 0)$, (a, a), $(-a, a)$. Using the auxiliary conic

$$x^2 - ay = tx\,(y - a),$$

we find $x = \dfrac{a}{t^3}(2 - 3t^2), \qquad y = \dfrac{a}{2t^4}(2 - 3t^2)(2 - t^2).$

(iii) (a) The curve $y^n = x^n + ax^{n-1}$

has a multiple point of order $n-1$ at the origin, and is therefore unicursal. In this case it is sufficient to consider the intersection of the curve with the line $y = tx$. This may be harmonised with the general theory by regarding the curve

$$y^{n-3}\,(y - tx) = 0,$$

as passing through each of the $\frac{1}{2}(n-1)(n-2)$ double points collected at the origin and through $n-3$ other fixed points collected at the point

$$x = -a, \quad y = 0.$$

The curves $y^n = x^n + ax^{n-1}$ ....(1),

$$y^n = 1 + az \;\; ....(2),$$

are projectively equivalent, as appears on rendering their equations homogeneous by the introduction of variables z in (1) and x in (2). We conclude that (2) is unicursal, having the maximum number of double points at infinity. In fact we may put

$$y = t, \quad az = t^n - 1.$$

The integral $\displaystyle\int R\,\{z,\; \sqrt[n]{(1 + az)}\}\, dz$

is accordingly an elementary function.

(b) The curve $y^m = A\,(x - a)^\mu\,(x - b)^\nu$

is unicursal if and only if either (i) $\mu = 0$ or (ii) $\nu = 0$ or (iii) $\mu + \nu = m$. Hence the integral

$$\int R\,\{x,\; (x - a)^{\mu/m}\,(x - b)^{\nu/n}\}\, dx$$

is an elementary function, for all forms of R, in these three cases only; of course it is integrable for special forms of R in other cases[*].

---

[*] See Ptaszycki, 'Extrait d'une lettre adressée à M. Hermite', *Bulletin des sciences mathématiques*, ser. 2, vol. 12, 1888, pp. 262–270: Appell and Goursat, *Théorie des fonctions algébriques*, p. 245.

9. There is a similar theory connected with unicursal curves in space of any number of dimensions. Consider for example the integral

$$\int R\left\{x,\ \sqrt{(ax+b)},\ \sqrt{(cx+d)}\right\} dx.$$

A linear substitution $x = lx + m$ reduces this integral to the form

$$\int R_1\left\{y,\ \sqrt{(y+2)},\ \sqrt{(y-2)}\right\} dy\ ;$$

and this integral can be rationalised by putting

$$y = t^2 + \frac{1}{t^2}\,,\quad \sqrt{(y+2)} = t + \frac{1}{t}\,,\quad \sqrt{(y-2)} = t - \frac{1}{t}\,.$$

The curve whose Cartesian coordinates $\xi,\ \eta,\ \zeta$ are given by

$$\xi : \eta : \zeta : 1 :: t^4 + 1 : t\,(t^2+1) : t\,(t^2-1) : t^2,$$

is a unicursal twisted quartic, the intersection of the parabolic cylinders

$$\xi = \eta^2 - 2,\quad \xi = \zeta^2 + 2.$$

It is easy to deduce that the integral

$$\int R\left\{x,\ \sqrt{\left(\frac{ax+b}{mx+n}\right)},\ \sqrt{\left(\frac{cx+d}{mx+n}\right)}\right\} dx$$

is always an elementary function.

10. When the deficiency of the curve $f(x,\ y) = 0$ is not zero, the integral

$$\int R\,(x,\ y)\,dx$$

is in general not an elementary function; and the consideration of such integrals has consequently introduced a whole series of classes of new transcendents into analysis. The simplest case is that in which the deficiency is unity: in this case, as we shall see later on, the integrals are expressible in terms of elementary functions and certain new transcendents known as elliptic integrals. When the deficiency rises above unity the integration necessitates the introduction of new transcendents of growing complexity.

But there are infinitely many particular cases in which integrals, associated with curves whose deficiency is unity or greater than unity, can be expressed in terms of elementary functions, or are even algebraical themselves. For instance the deficiency of

$$y^2 = 1 + x^3$$

is unity. But

$$\int \frac{x+1}{x-2} \frac{dx}{\sqrt{(1+x^3)}} = 3 \log \frac{(1+x)^2 - 3\sqrt{(1+x^3)}}{(1+x)^2 + 3\sqrt{(1+x^3)}},$$

$$\int \frac{2-x^3}{1+x^3} \frac{dx}{\sqrt{(1+x^3)}} = \frac{2x}{\sqrt{(1+x^3)}}.$$

And, before we say anything concerning the new transcendents to which integrals of this class in general give rise, we shall consider what has been done in the way of formulating rules to enable us to identify such cases and to assign the form of the integral when it is an elementary function. It will be as well to say at once that this problem has not been solved completely.

11. The first general theorem of this character deals with the case in which the integral is algebraical, and asserts that *if*

$$u = \int y \, dx$$

is an algebraical function of x, then it is a rational function of x and y.

Our proof will be based on the following lemmas.

(1) *If $f(x, y)$ and $g(x, y)$ are polynomials, and there is no factor common to all the coefficients of the various powers of y in $g(x, y)$; and*

$$f(x, y) = g(x, y) h(x),$$

where $h(x)$ is a rational function of x; then $h(x)$ is a polynomial.

Let $h = P/Q$, where P and Q are polynomials without a common factor. Then

$$fQ = gP.$$

If $x - a$ is a factor of Q, then

$$g(a, y) = 0$$

for all values of y; and so all the coefficients of powers of y in $g(x, y)$ are divisible by $x - a$, which is contrary to our hypotheses. Hence Q is a constant and h a polynomial.

(2) *Suppose that $f(x, y)$ is an irreducible polynomial, and that $y_1, y_2, \ldots, y_n$ are the roots of*

$$f(x, y) = 0$$

in a certain domain D. Suppose further that $\phi(x, y)$ is another polynomial, and that

$$\phi(x, y_1) = 0.$$

Then $\qquad \phi(x, y_s) = 0,$

where y_s is any one of the roots of (1); and

$$\phi(x, y) = f(x, y)\psi(x, y),$$

where $\psi(x, y)$ also is a polynomial in x and y.

Let us determine the highest common factor ϖ of f and ϕ, considered as polynomials in y, by the ordinary process for the determination of the highest common factor of two polynomials. This process depends only on a series of algebraical divisions, and so ϖ is a polynomial in y with coefficients rational in x. We have therefore

$$\varpi(x, y) = \omega(x, y)\lambda(x) \quad \ldots\ldots\ldots\ldots\ldots(1),$$

$$f(x, y) = \omega(x, y)p(x, y)\mu(x) = g(x, y)\mu(x) \quad \ldots\ldots(2),$$

$$\phi(x, y) = \omega(x, y)q(x, y)\nu(x) = h(x, y)\nu(x) \quad \ldots\ldots(3),$$

where ω, p, q, g, and h are polynomials and λ, μ, and ν rational functions; and evidently we may suppose that neither in g nor in h have the coefficients of all powers of y a common factor. Hence, by Lemma (1), μ and ν are polynomials. But f is irreducible, and therefore μ and either ω or p must be constants. If ω were a constant, ϖ would be a function of x only. But this is impossible. For we can determine polynomials L, M in y, with coefficients rational in x, such that

$$Lf + M\phi = \varpi \quad \ldots\ldots\ldots\ldots\ldots\ldots\ldots(4),$$

and the left-hand side of (4) vanishes when we write y_1 for y. Hence p is a constant, and so ω is a constant multiple of f. The truth of the lemma now follows from (3).

It follows from Lemma (2) that *y cannot satisfy any equation of degree less than n whose coefficients are polynomials in x.*

(3) *If y is an algebraical function of x, defined by an equation*

$$f(x, y) = 0 \quad \ldots\ldots\ldots\ldots\ldots\ldots\ldots(1)$$

of degree n, then any rational function $R(x, y)$ of x and y can be expressed in the form

$$R(x, y) = R_0 + R_1 y + \ldots + R_{n-1}y^{n-1} \quad \ldots\ldots\ldots(2),$$

where R_0, R_1, ..., R_{n-1} are rational functions of x.

The function y is one of the n roots of (1). Let $y, y', y'', \ldots$ be the complete system of roots.   Then

$$R(x, y) = \frac{P(x, y)}{Q(x, y)}$$

$$= \frac{P(x, y)\, Q(x, y')\, Q(x, y'') \cdots}{Q(x, y)\, Q(x, y')\, Q(x, y'') \cdots} \quad \ldots\ldots\ldots\ldots(3),$$

where P and Q are polynomials.   The denominator is a polynomial in x whose coefficients are symmetric polynomials in $y, y', y'', \ldots$, and is therefore, by II., § 3, (i), a rational function of x.   On the other hand

$$Q(x, y')\, Q(x, y'') \cdots$$

is a polynomial in x whose coefficients are symmetric polynomials in $y', y'', \ldots$, and therefore, by II., § 3, (ii), polynomials in y with coefficients rational in x.   Thus the numerator of (3) is a polynomial in y with coefficients rational in x.

It follows that $R(x, y)$ is a polynomial in y with coefficients rational in x.   From this polynomial we can eliminate, by means of (1), all powers of y as high as or higher than the nth.   Hence $R(x, y)$ is of the form prescribed by the lemma.

12.   We proceed now to the proof of our main theorem.   We have

$$\int y\, dx = u$$

where u is algebraical.   Let

$$f(x, y) = 0, \quad \psi(x, u) = 0 \quad \ldots\ldots\ldots\ldots\ldots\ldots(1)$$

be the irreducible equations satisfied by y and u, and let us suppose that they are of degrees n and m respectively.   The first stage in the proof consists in showing that

$$m = n.$$

It will be convenient now to write y_1, u_1 for y, u, and to denote by

$$y_1, y_2, \ldots, y_n, \quad u_1, u_2, \ldots, u_m,$$

the complete systems of roots of the equations (1).

We have

$$\psi(x, u_1) = 0,$$

and so

$$\chi_1 = \frac{\partial \psi}{\partial x} + \frac{\partial \psi}{\partial u_1}\frac{du_1}{dx} = \frac{\partial \psi}{\partial x} + y_1 \frac{\partial \psi}{\partial u_1} = 0.$$

Now let

$$\Omega(x, u_1) = \prod_{r=1}^{n} \left(\frac{\partial \psi}{\partial x} + y_r \frac{\partial \psi}{\partial u_1}\right).$$

Then Ω is a polynomial in u_1, with coefficients symmetric in $y_1, y_2, \ldots, y_n$ and therefore rational in x.

The equations $\psi = 0$ and $\Omega = 0$ have a root u_1 in common, and the first equation is irreducible. It follows, by Lemma (2) of § 11, that

$$\Omega\,(x, u_s) = 0$$

for $s = 1, 2, \ldots, m.^*$ And from this it follows that, when s is given, we have

$$\frac{\partial\psi}{\partial x} + y_r\,\frac{\partial\psi}{\partial u_s} = 0 \dots\dots\dots\dots\dots\dots\dots(2)$$

for some value of the suffix r.

But we have also

$$\frac{\partial\psi}{\partial x} + \frac{\partial\psi}{\partial u_s}\,\frac{du_s}{dx} = 0 \dots\dots\dots\dots\dots\dots(3);$$

and from (2) and (3) it follows † that

$$\frac{du_s}{dx} = y_r \quad \dots\dots\dots\dots\dots\dots\dots(4),$$

i.e. that *every u is the integral of some y.*

In the same way we can show that *every y is the derivative of some u.* Let

$$\omega\,(x, y_1) = \prod_{s=1}^{m}\,\left(\frac{\partial\psi}{\partial x} + y_1\,\frac{\partial\psi}{\partial u_s}\right).$$

Then ω is a polynomial in y_1, with coefficients symmetric in $u_1, u_2, \ldots, u_m$ and therefore rational in x. The equations $f = 0$ and $\omega = 0$ have a root y_1 in common, and so

$$\omega\,(x, y_r) = 0$$

for $r = 1, 2, \ldots, n$. From this we deduce that, when r is given, (2) must be true for some value of s, and so that the same is true of (4).

Now it is impossible that, in (4), two different values of s should correspond to the same value of r. For this would involve

$$u_s - u_t = c$$

where $s \neq t$ and c is a constant. Hence we should have

$$\psi\,(x, u_s) = 0, \quad \psi\,(x, u_s - c) = 0.$$

---

* If $p\,(x)$ is the least common multiple of the denominators of the coefficients of powers of u in Ω, then

$$\Omega\,(x, u)\,p\,(x) = \chi\,(x, u),$$

where χ is a polynomial. Applying Lemma (2), we see that $\chi\,(x, u_s) = 0$, and so

$$\Omega\,(x, u_s) = 0.$$

† It is impossible that ψ and $\dfrac{\partial\psi}{\partial u}$ should both vanish for $u = u_s$, since ψ is irreducible.

Subtracting these equations, we should obtain an equation of degree $m-1$ in u_s, with coefficients which are polynomials in x; and this is impossible. In the same way we can prove that two different values of r cannot correspond to the same value of s.

The equation (4) therefore establishes a one-one correspondence between the values of r and s. It follows that

$$m = n.$$

It is moreover evident that, by arranging the suffixes properly, we can make

$$\frac{du_r}{dx} = y_r \quad\dots\dots\dots\dots\dots\dots\dots\dots\dots(5)$$

for $r = 1, 2, \dots, n$.

13. We have

$$y_r = \frac{du_r}{dx} = -\frac{\partial \psi}{\partial x} \Big/ \frac{\partial \psi}{\partial u_r} = R(x, u_r),$$

where R is a rational function which may, in virtue of Lemma (3) of §11, be expressed as a polynomial of degree $n-1$ in u_r, with coefficients rational in x.

The product

$$\underset{s \neq r}{\Pi}\,(z - y_s)$$

is a polynomial of degree $n-1$ in z, with coefficients which are symmetric polynomials in $y_1,\ y_2,\ \dots,\ y_{r-1},\ y_{r+1},\ \dots,\ y_n$ and therefore, by II., §3, (ii), polynomials in y_r with coefficients rational in x. Replacing y_r by its expression as a polynomial in u_r obtained above, and eliminating u_r^n and all higher powers of u_r, we obtain an equation

$$\underset{s \neq r}{\Pi}\,(z - y_s) = \sum_{j=0}^{n-1} \sum_{k=0}^{n-1} S_{j,k}(x)\, z^j u_r^k,$$

where the S's are rational functions of x which are, from the method of their formation, independent of the particular value of r selected. We may therefore write

$$\underset{s \neq r}{\Pi}\,(z - y_s) = P(x, z, u_r),$$

where P is a polynomial in z and u_r with coefficients rational in x. It is evident that

$$P(x, y_s, u_r) = 0$$

for every value of s other than r. In particular

$$P(x, y_1, u_r) = 0 \qquad\qquad (r = 2, 3, \dots,$$

It follows that the $n-1$ roots of the equation in u

$$P(x, y_1, u) = 0$$

are $u_2, u_3, \ldots, u_n$.   We have therefore

$$P(x, y_1, u) = T_0(x, y_1) \prod_2^n (u - u_r)$$

$$= T_0(x, y_1) \{u^{n-1} - u^{n-2}(u_2 + u_3 + \ldots + u_n) + \ldots\}$$

$$= T_0(x, y_1) \left[u^{n-1} + u^{n-2} \left\{ u_1 + \frac{B_1(x)}{B_0(x)} \right\} + \ldots \right],$$

where $T_0(x, y_1)$ is the coefficient of u^{n-1} in P, and $B_0(x)$ and $B_1(x)$ are the coefficients of u^n and u^{n-1} in ψ.   Equating the coefficients of u^{n-2} on the two sides of this equation, we obtain

$$u_1 + \frac{B_1(x)}{B_0(x)} = \frac{T_1(x, y_1)}{T_0(x, y_1)},$$

where $T_1(x, y_1)$ is the coefficient of u^{n-2} in P.   Thus the theorem is proved.

14.   We can now apply Lemma (3) of § 11 ; and we arrive at the final conclusion that *if*

$$\int y\, dx$$

is algebraical then it can be expressed in the form

$$R_0 + R_1 y + \ldots + R_{n-1} y^{n-1},$$

where $R_0, R_1, \ldots$ are rational functions of x.

The most important case is that in which

$$y = \sqrt[n]{\{R(x)\}},$$

where $R(x)$ is rational.   In this case

$$y^n = R(x) \quad\ldots\ldots\ldots\ldots\ldots\ldots\ldots\ldots\ldots(1),$$

$$\frac{dy}{dx} = \frac{R'(x)}{n y^{n-1}} \quad\ldots\ldots\ldots\ldots\ldots\ldots\ldots\ldots(2).$$

But

$$y = R_0' + R_1' y + \ldots + R'_{n-1} y^{n-1}$$

$$+ \{R_1 + 2R_2 y + \ldots + (n-1)R_{n-1}y^{n-2}\} \frac{dy}{dx} \ldots\ldots\ldots\ldots(3).$$

Eliminating $\dfrac{dy}{dx}$ between these equations, we obtain an equation

$$\varpi(x, y) = 0 \ldots\ldots\ldots\ldots\ldots\ldots\ldots\ldots\ldots(4),$$

where $\varpi(x, y)$ is a polynomial.   It follows from Lemma (2) of § 11 that this equation must be satisfied by all the roots of (1).   Thus (4) is still true if we replace y by any other root y' of (1) ; and as

(2) is still true when we effect this substitution, it follows that (3) is also still true. Integrating, we see that the equation

$$\int y\,dx = R_0 + R_1 y + \dots + R_{n-1}y^{n-1}$$

is true when y is replaced by y'. We may therefore replace y by ωy, ω being any primitive nth root of unity. Making this substitution, and multiplying by ω^{n-1}, we obtain

$$\int y\,dx = \omega^{n-1}R_0 + R_1 y + \omega R_2 y + \dots + \omega^{n-2}R_{n-1}y^{n-1}\,;$$

and on adding the n equations of this type we obtain

$$\int y\,dx = R_1 y.$$

Thus in this case the functions R_0, R_2, ..., R_{n-1} all disappear.

It has been shown by Liouville* that the preceding results enable us to obtain in all cases, by a finite number of elementary algebraical operations, a solution of the problem '*to determine whether $\int y\,dx$ is algebraical, and to find the integral when it is algebraical*'.

15. It would take too long to attempt to trace in detail the steps of the general argument. We shall confine ourselves to a solution of a particular problem which will give a sufficient illustration of the general nature of the arguments which must be employed.

We shall determine under what circumstances the integral

$$\int \frac{dx}{(x-p)\,\surd(ax^2+2bx+c)}$$

is algebraical. This question might of course be answered by actually evaluating the integral in the general case and finding when the integral function reduces to an algebraical function. We are now, however, in a position to answer it without any such integration.

We shall suppose first that $ax^2+2bx+c$ is not a perfect square. In this case

$$y = \frac{1}{\surd X},$$

where

$$X = (x-p)^2\,(ax^2+2bx+c),$$

and if $\int y\,dx$ is algebraical it must be of the form

$$\frac{R(x)}{\surd X}\,.$$

Hence

$$y = \frac{d}{dx}\left(\frac{R}{\surd X}\right),$$

or

$$2X = 2XR' - RX'.$$

* 'Premier mémoire sur la détermination des intégrales dont la valeur est algébrique', *Journal de l'École Polytechnique*, vol. 14, cahier 22, 1833, pp. 124–148 ; 'Second mémoire...', *ibid.*, pp. 149–193.

We can now show that R is a polynomial in x. For if $R = U/V$, where U and V are polynomials, then V, if not a mere constant, must contain a factor

$$(x-a)^\mu \qquad\qquad (\mu > 0),$$

and we can put
$$R = \frac{U}{W(x-a)^\mu},$$

where U and W do not contain the factor $x - a$. Substituting this expression for R, and reducing, we obtain

$$\frac{2\mu\, U W X}{x-a} = 2U'WX - 2UW'X - UWX' - 2W^2X(x-a)^\mu.$$

Hence X must be divisible by $x - a$. Suppose then that

$$X = (x-a)^k\, Y,$$

where Y is prime to $x - a$. Substituting in the equation last obtained we deduce

$$\frac{(2\mu + k)\, U W Y}{x-a} = 2U'WY - 2UW'Y - UWY' - 2W^2Y(x-a)^\mu,$$

which is obviously impossible, since neither U, W, nor Y is divisible by $x - a$. Thus V must be a constant. Hence

$$\int \frac{dx}{(x-p)\sqrt{(ax^2+2bx+c)}} = \frac{U(x)}{(x-p)\sqrt{(ax^2+2bx+c)}},$$

where $U(x)$ is a polynomial.

Differentiating and clearing of radicals we obtain

$$\{(x-p)(U'-1) - U\}(ax^2+2bx+c) = U(x-p)(ax+b).$$

Suppose that the first term in U is Ax^m. Equating the coefficients of x^{m+2}, we find at once that $m = 2$. We may therefore take

$$U = Ax^2 + 2Bx + C,$$

so that

$$\{(x-p)(2Ax+2B-1) - Ax^2 - 2Bx - C\}(ax^2+2bx+c)$$
$$= (x-p)(ax+b)(Ax^2+2Bx+C)......(1).$$

From (1) it follows that

$$(x-p)(ax+b)(Ax^2+2Bx+C)$$

is divisible by $ax^2+2bx+c$. But $ax+b$ is not a factor of $ax^2+2bx+c$, as the latter is not a perfect square. Hence either (i) $ax^2+2bx+c$ and $Ax^2+2Bx+C$ differ only by a constant factor or (ii) the two quadratics have one and only one factor in common, and $x - p$ is also a factor of $ax^2+2bx+c$. In the latter case we may write

$$ax^2+2bx+c = a(x-p)(x-q), \quad Ax^2+2Bx+C = A(x-q)(x-r),$$

where $p \neq q$, $p \neq r$. It then follows from (1) that

$$a(x-p)(2Ax+2B-1) - aA(x-q)(x-r) = A(ax+b)(x-r).$$

Hence $2Ax+2B-1$ is divisible by $x - r$. Dividing by $aA(x-r)$ we obtain

$$2(x-p) - (x-q) = x + \frac{b}{a} = x - \tfrac{1}{2}(p+q),$$

and so $p = q$, which is untrue.

Hence case (ii) is impossible, and so $ax^2+2bx+c$ and $Ax^2+2Bx+C$ differ only by a constant factor. It then follows from (1) that $x-p$ is a factor of $ax^2+2bx+c$; and the result becomes

$$\int \frac{dx}{(x-p)\sqrt{(ax^2+2bx+c)}} = K\frac{\sqrt{(ax^2+2bx+c)}}{x-p},$$

where K is a constant. It is easily verified that this equation is actually true when $ap^2+2bp+c=0$, and that

$$K=\frac{1}{\sqrt{(b^2-ac)}}.$$

The formula is equivalent to

$$\int \frac{dx}{(x-p)\sqrt{\{(x-p)(x-q)\}}} = \frac{2}{q-p}\sqrt{\left(\frac{x-q}{x-p}\right)}.$$

There remains for consideration the case in which $ax^2+2bx+c$ is a perfect square, say $a(x-q)^2$. Then

$$\int \frac{dx}{(x-p)(x-q)}$$

must be rational, and so $p=q$.

As a further example, the reader may verify that if

$$y^3-3y+2x=0$$

then

$$\int y\,dx = \frac{3}{8}(2xy-y^2).*$$

16.   The theorem of § 11 enables us to complete the proof of the two fundamental theorems stated without proof in II., § 5, viz.

(a) e^x is not an algebraical function of x,

(b) $\log x$ is not an algebraical function of x.

We shall prove (b) as a special case of a more general theorem, viz. ' *no sum of the form*

$$A\log(x-a)+B\log(x-\beta)+\dots,$$

in which the coefficients A, B, ... are not all zero, can be an algebraical function of x '. To prove this we have only to observe that the sum in question is the integral of a rational function of x. If then it is algebraical it must, by the theorem of § 11, be rational, and this we have already seen to be impossible (IV., 2).

That e^x is not algebraical now follows at once from the fact that it is the inverse function of $\log x$.

17.   The general theorem of § 11 gives the first step in the rigid proof of 'Laplace's principle' stated in III., § 2. On account of the immense importance of this principle we repeat Laplace's words:

* Raffy, ' Sur les quadratures algébriques et logarithmiques ', *Annales de l'École Normale*, ser. 3, vol. 2, 1885, pp. 185–206.

'*l'intégrale d'une fonction différentielle ne peut contenir d'autres quantités radicaux que celles qui entrent dans cette fonction*'.   This general principle, combined with arguments similar to those used above (§ 15) in a particular case, enables us to prove without difficulty that a great many integrals cannot be algebraical, notably the standard elliptic integrals

$$\int \frac{dx}{\sqrt{\{(1-x^2)(1-k^2x^2)\}}}, \quad \int \sqrt{\left(\frac{1-x^2}{1-k^2x^2}\right)} \, dx, \quad \int \frac{dx}{\sqrt{(4x^3 - g_2 x - g_3)}}$$

which give rise by inversion to the elliptic functions.

18.   We must now consider in a very summary manner the more difficult question of the nature of those integrals of algebraical functions which are expressible in finite terms by means of the elementary transcendental functions.   In the first place *no integral of any algebraical function can contain any exponential*.   Of this theorem it is, as we remarked before, easy to become convinced by a little reflection, as doubtless did Laplace, who certainly possessed no rigorous proof. The reader will find little difficulty in coming to the conclusion that exponentials cannot be eliminated from an elementary function by differentiation.   But we would strongly recommend him to study the exceedingly beautiful and ingenious proof of this proposition given by Liouville*.   We have unfortunately no space to insert it here.

It is instructive to consider particular cases of this theorem. Suppose for example that $\int y\,dx$, where y is algebraical, were a polynomial in x and e^x, say

$$\Sigma\Sigma a_{m,n} x^m e^{nx} \quad \dots\dots\dots\dots\dots\dots\dots\dots\dots\dots(1).$$

When this expression is differentiated, e^x must disappear from it: otherwise we should have an algebraical relation between x and e^x. Expressing the conditions that the coefficient of every power of e^x in the differential coefficient of (1) vanishes identically, we find that the same must be true of (1), so that after all the integral does not really contain e^x. Liouville's proof is in reality a development of this idea.

The integral of an algebraical function, if expressible in terms of elementary functions, can therefore only contain algebraical or logarithmic functions.   The next step is to show that the logarithms must be simple logarithms of algebraical functions and can only enter linearly, so that the general integral must be of the type

$$\int y\,dx = u + A \log v + B \log w + \dots,$$

---

* '*Mémoire sur les transcendantes elliptiques considérées comme fonctions de leur amplitude*', *Journal de l'École Polytechnique*, vol. 14, cahier 23, 1834, pp. 37–83. The proof may also be found in Bertrand's *Calcul intégral*, p. 99.

where A, B, ... are constants and u, v, w, ... algebraical functions. Only when the logarithms occur in this simple form will differentiation eliminate them.

Lastly it can be shown by arguments similar to those of §§ 11–14 that u, v, w, ... are rational functions of x and y. Thus $\int y\,dx$, if an elementary function, is *the sum of a rational function of x and y and of certain constant multiples of logarithms of such functions.* We can suppose that no two of A, B, ... are commensurable, or indeed, more generally, that no linear relation

$$A\alpha + B\beta + \ldots = 0,$$

with rational coefficients, holds between them. For if such a relation held then we could eliminate A from the integral, writing it in the form

$$\int y\,dx = u + B \log \left(wv^{-\beta/\alpha}\right) + \ldots$$

It is instructive to verify the truth of this theorem in the special case in which the curve $f(x, y) = 0$ is unicursal. In this case x and y are rational functions $R(t)$, $S(t)$ of a parameter t, and the integral, being the integral of a rational function of t, is of the form

$$u + A \log v + B \log w + \ldots,$$

where u, v, w, ... are rational functions of t. But t may be expressed, by means of elementary algebraical operations, as a rational function of x and y. Thus u, v, w, ... are rational functions of x and y.

The case of greatest interest is that in which y is a rational function of x and $\sqrt{X}$, where X is a polynomial. As we have already seen, y can in this case be expressed in the form

$$P + \frac{Q}{\sqrt{X}},$$

where P and Q are rational functions of x. We shall suppress the rational part and suppose that $y = Q/\sqrt{X}$. In this case the general theorem gives

$$\int \frac{Q}{\sqrt{X}}\,dx = S + \frac{T}{\sqrt{X}} + A \log\left(\alpha + \beta\sqrt{X}\right) + B \log\left(\gamma + \delta\sqrt{X}\right) + \ldots,$$

where S, T, α, β, γ, δ, ... are rational. If we differentiate this equation we obtain an algebraical identity in which we can change the sign of $\sqrt{X}$. Thus we may change the sign of $\sqrt{X}$ in the integral equation. If we do this and subtract, and write $2A, \ldots$ for $A, \ldots$, we obtain

$$\int \frac{Q}{\sqrt{X}}\,dx = \frac{T}{\sqrt{X}} + A \log \frac{\alpha + \beta\sqrt{X}}{\alpha - \beta\sqrt{X}} + B \log \frac{\gamma + \delta\sqrt{X}}{\gamma - \delta\sqrt{X}} + \ldots,$$

which is the standard form for such an integral.   It is evident that we may suppose $a, \beta, \gamma, \ldots$ to be polynomials.

19. (i) By means of this theorem it is possible to prove that a number of important integrals, and notably the integrals

$$\int \frac{dx}{\sqrt{\{(1-x^2)(1-k^2x^2)\}}}, \qquad \int \sqrt{\left\{\frac{1-x^2}{1-k^2x^2}\right\}}\, dx, \qquad \int \frac{dx}{\sqrt{(4x^3-g_2x-g_3)}},$$

are not expressible in terms of elementary functions, and so represent genuinely new transcendents.   The formal proof of this was worked out by Liouville*; it rests merely on a consideration of the possible forms of the differential coefficients of expressions of the form

$$\frac{T}{\sqrt{X}} + A \log \frac{a+\beta\sqrt{X}}{a-\beta\sqrt{X}} + \ldots,$$

and the arguments used are purely algebraical and of no great theoretical difficulty.   The proof is however too detailed to be inserted here.   It is not difficult to find shorter proofs, but these are of a less elementary character, being based on ideas drawn from the theory of functions†.

The general questions of this nature which arise in connection with integrals of the form

$$\int \frac{Q}{\sqrt{X}}\, dx,$$

or, more generally,

$$\int \frac{Q}{\sqrt[m]{X}}\, dx,$$

are of extreme interest and difficulty.   The case which has received most attention is that in which $m=2$ and X is of the third or fourth degree, in which case the integral is said to be *elliptic*.   An integral of this kind is called *pseudo-elliptic* if it is expressible in terms of algebraical and logarithmic functions.   Two examples were given above (§ 10).   General methods have been given for the construction of such integrals, and it has been shown that certain interesting forms are pseudo-elliptic.   In Goursat's *Cours d'analyse*‡, for instance, it is shown that if $f(x)$ is a rational function such that

$$f(x)+f\left(\frac{1}{k^2x}\right)=0,$$

then

$$\int \frac{f(x)\, dx}{\sqrt{\{x(1-x)(1-k^2x)\}}}$$

is pseudo-elliptic.   But no method has been devised as yet by which we can always determine in a finite number of steps whether a *given* elliptic integral

---

* See Liouville's memoir quoted on p. 45 (pp. 45 *et seq.*).

† The proof given by Laurent (*Traité d'analyse*, vol. 4, pp. 153 *et seq.*) appears at first sight to combine the advantages of both methods of proof, but unfortunately will not bear a closer examination.

‡ Second edition, vol. 1, pp. 267-269.

is pseudo-elliptic, and integrate it if it is, and there is reason to suppose that no such method can be given. And up to the present it has not, so far as we know, been proved rigorously and explicitly that (*e.g.*) the function

$$u = \int \frac{dx}{\sqrt{\{(1-x^2)(1-k^2x^2)\}}}$$

is not a root of an elementary transcendental equation; all that has been shown is that it is not *explicitly* expressible in terms of elementary transcendents. The processes of reasoning employed here, and in the memoirs to which we have referred, do not therefore suffice to prove that the inverse function $x = \operatorname{sn} u$ is not an elementary function of u. Such a proof must rest on the known properties of the function $\operatorname{sn} u$, and would lie altogether outside the province of this tract.

The reader who desires to pursue the subject further will find references to the original authorities in Appendix I.

(ii) One particular class of integrals which is of especial interest is that of the *binomial integrals*

$$\int x^m (ax^n + b)^p \, dx,$$

where m, n, p are rational. Putting $ax^n = bt$, and neglecting a constant factor, we obtain an integral of the form

$$\int t^q (1+t)^p \, dt,$$

where p and q are rational. If p is an integer, and q a fraction r/s, this integral can be evaluated at once by putting $t = u^s$, a substitution which rationalises the integrand. If q is an integer, and $p = r/s$, we put $1 + t = u^s$. If $p + q$ is an integer, and $p = r/s$, we put $1 + t = tu^s$.

It follows from Tschebyschef's researches (to which references are given in Appendix I) that these three cases are the only ones in which the integral can be evaluated in finite form.

20.   In §§ 7–9 we considered in some detail the integrals connected with curves whose deficiency is zero. We shall now consider in a more summary way the case next in simplicity, that in which the deficiency is unity, so that the number of double points is

$$\tfrac{1}{2}(n-1)(n-2) - 1 = \tfrac{1}{2}n(n-3).$$

It has been shown by Clebsch* that in this case the coordinates of the points of the curve can be expressed as *rational functions of a parameter t and of the square root of a polynomial in t of the third or fourth degree.*

* 'Über diejenigen Curven, deren Coordinaten sich als elliptische Functionen eines Parameters darstellen lassen', *Journal für Mathematik*, vol. 64, 1865, pp. 210–270.

The fact is that the curves

$$y^2 = a + bx + cx^2 + dx^3,$$
$$y^2 = a + bx + cx^2 + dx^3 + ex^4,$$

are the simplest curves of deficiency 1. The first is the typical cubic without a double point. The second is a quartic with two double points, in this case coinciding in a 'tacnode' at infinity, as we see by making the equation homogeneous with z, writing 1 for y, and then comparing the resulting equation with the form treated by Salmon on p. 215 of his *Higher plane curves*. The reader who is familiar with the theory of algebraical plane curves will remember that the deficiency of a curve is unaltered by any birational transformation of coordinates, and that any curve can be birationally transformed into any other curve of the same deficiency, so that any curve of deficiency 1 can be birationally transformed into the cubic whose equation is written above.

The argument by which this general theorem is proved is very much like that by which we proved the corresponding theorem for unicursal curves. The simplest case is that of the general cubic curve. We take a point on the curve as origin, so that the equation of the curve is of the form

$$ax^3 + 3bx^2y + 3cxy^2 + dy^3 + ex^2 + 2fxy + gy^2 + hx + ky = 0.$$

Let us consider the intersections of this curve with the secant $y = tx$. Eliminating y, and solving the resulting quadratic in x, we see that the only irrationality which enters into the expression of x is

$$\sqrt{(T_2^2 - 4T_1 T_3)},$$

where $\quad T_1 = h + kt, \quad T_2 = e + 2ft + gt^2, \quad T_3 = a + 3bt + 3ct^2 + dt^3.$

A more elegant method has been given by Clebsch[*]. If we write the cubic in the form

$$LMN = P,$$

where L, M, N, P are linear functions of x and y, so that L, M, N are the asymptotes, then the hyperbolas $LM = t$ will meet the cubic in four fixed points at infinity, and therefore in two points only which depend on t. For these points

$$LM = t, \quad P = tN.$$

Eliminating y from these equations, we obtain an equation of the form

$$Ax^2 + 2Bx + C = 0,$$

where A, B, C are quadratics in t. Hence

$$x = -\frac{B}{A} \pm \frac{\sqrt{(B^2 - AC)}}{A} = R(t, \sqrt{T}),$$

---

[*] See Hermite, *Cours d'analyse*, pp. 422–425.

where $T = B^2 - AC$ is a polynomial in t of degree not higher than the fourth.

Thus if the curve is

$$x^3 + y^3 - 3axy + 1 = 0,$$

so that

$$L = \omega x + \omega^2 y + a, \quad M = \omega^2 x + \omega y + a, \quad N = x + y + a, \quad P = a^3 - 1,$$

ω being an imaginary cube root of unity, then we find that the line

$$x + y + a = \frac{a^3 - 1}{t}$$

meets the curve in the points given by

$$x = \frac{b - at}{2t} \pm \frac{\sqrt{(3T)}}{6t}, \qquad y = \frac{b - at}{2t} \mp \frac{\sqrt{(3T)}}{6t},$$

where $b = a^3 - 1$ and

$$T = 4t^3 - 9a^2 t^2 + 6abt - b^2.$$

In particular, for the curve

$$x^3 + y^3 + 1 = 0,$$

we have

$$x = \frac{-\sqrt{3} + \sqrt{(4t^3 - 1)}}{2t\sqrt{3}}, \qquad y = \frac{-\sqrt{3} - \sqrt{(4t^3 - 1)}}{2t\sqrt{3}}.$$

21. It will be plain from what precedes that

$$\int R\{x, \sqrt[3]{(a + bx + cx^2 + dx^3)}\}\, dx$$

can always be reduced to an elliptic integral, the deficiency of the cubic

$$y^3 = a + bx + cx^2 + dx^3$$

being unity.

In general integrals associated with curves whose deficiency is greater than unity cannot be so reduced.   But associated with every curve of, let us say, deficiency 2 there will be an infinity of integrals

$$\int R(x, y)\, dx$$

reducible to elliptic integrals or even to elementary functions ; and there are curves of deficiency 2 for which *all* such integrals are reducible.

For example, the integral

$$\int R\{x, \sqrt{(x^6 + ax^4 + bx^2 + c)}\}\, dx$$

may be split up into the sum of the integral of a rational function and two integrals of the types

$$\int \frac{R\,(x^2)\,dx}{\sqrt{(x^6 + ax^4 + bx^2 + c)}}, \qquad \int \frac{xR\,(x^2)\,dx}{\sqrt{(x^6 + ax^4 + bx^2 + c)}},$$

and each of these integrals becomes elliptic on putting $x^2 = t$. But the deficiency of

$$y^2 = x^6 + ax^4 + bx^2 + c$$

is 2. Another example is given by the integral

$$\int R\,\{x,\ \sqrt[3]{(x^4 + ax^3 + bx^2 + cx + d)}\}\,dx.*$$

22. It would be beside our present purpose to enter into any details as to the general theory of elliptic integrals, still less of the integrals (usually called Abelian) associated with curves of deficiency greater than unity. We have seen that if the deficiency is unity then the integral can be transformed into the form

$$\int R\,(x,\ \sqrt{X})\,dx$$

where $\qquad X = x^4 + ax^3 + bx^2 + cx + d.\dagger$

It can be shown that, by a transformation of the type

$$x = \frac{at + \beta}{\gamma t + \delta},$$

this integral can be transformed into an integral

$$\int R\,(t,\ \sqrt{T})\,dt$$

where $\qquad T = t^4 + A\,t^2 + B.$

We can then, as when T is of the second degree (§ 3), decompose this integral into two integrals of the forms

$$\int R\,(t)\,dt, \qquad \int \frac{R\,(t)\,dt}{\sqrt{T}}.$$

Of these integrals the first is elementary, and the second can be

---

* See Legendre, *Traité des fonctions elliptiques*, vol., 1, chs. 26–27, 32–33 ; Bertrand, *Calcul intégral*, pp. 67 *et seq.* ; and Enneper, *Elliptische Funktionen*, note 1, where abundant references are given.

† There is a similar theory for curves of deficiency 2, in which X is of the *sixth* degree.

decomposed* into the sum of an algebraical term, of certain multiples of the integrals

$$\int\frac{dt}{\sqrt{T}}, \qquad \int\frac{t^2 dt}{\sqrt{T}},$$

and of a number of integrals of the type

$$\int\frac{dt}{(t-\tau)\sqrt{T}}.$$

These integrals cannot in general be reduced to elementary functions, and are therefore new transcendents.

We will only add, before leaving this part of our subject, that the algebraical part of these integrals can be found by means of the elementary algebraical operations, as was the case with the rational part of the integral of a rational function, and with the algebraical part of the simple integrals considered in §§ 14–15.

VI. Transcendental functions

1. The theory of the integration of transcendental functions is naturally much less complete than that of the integration of rational or even of algebraical functions. It is obvious from the nature of the case that this must be so, as there is no general theorem concerning transcendental functions which in any way corresponds to the theorem that any algebraical combination of algebraical functions may be regarded as a simple algebraical function, the root of an equation of a simple standard type.

It is indeed almost true to say that there is no general theory, or that the theory reduces to an enumeration of the few cases in which the integral may be transformed by an appropriate substitution into an integral of a rational or algebraical function. These few cases are however of great importance in applications.

2. (i) The integral

$$\int F(e^{ax}, e^{bx}, \ldots, e^{kx})\, dx$$

where F is an algebraical function, and $a, b, \ldots, k$ commensurable numbers, can always be reduced to that of an algebraical function. In particular the integral

$$\int R(e^{ax}, e^{bx}, \ldots, e^{kx})\, dx,$$

* See, e.g., Goursat, Cours d'analyse, ed. 2, vol. 1, pp. 257 et seq.

where R is rational, is always an elementary function. In the first place a substitution of the type $x = ay$ will reduce it to the form

$$\int R\left(e^{y}\right) dy,$$

and then the substitution $e^{y} = z$ will reduce this integral to the integral of a rational function.

In particular, since $\cosh x$ and $\sinh x$ are rational functions of e^{x}, and $\cos x$ and $\sin x$ are rational functions of e^{ix}, the integrals

$$\int R\left(\cosh x,\ \sinh x\right) dx, \quad \int R\left(\cos x,\ \sin x\right) dx$$

are always elementary functions. In the second place the substitution just indicated is imaginary, and it is generally more convenient to use the substitution

$$\tan \tfrac{1}{2}x = t,$$

which reduces the integral to that of a rational function, since

$$\cos x = \frac{1 - t^{2}}{1 + t^{2}}, \quad \sin x = \frac{2t}{1 + t^{2}}, \quad dx = \frac{2dt}{1 + t^{2}}.$$

(ii)   The integrals

$$\int R\left(\cosh x,\ \sinh x,\ \cosh 2x,\ \ldots\ldots \sinh mx\right) dx,$$

$$\int R\left(\cos x,\ \sin x,\ \cos 2x,\ \ldots\ldots \sin mx\right) dx,$$

are included in the two standard integrals above.

Let us consider some further developments concerning the integral

$$\int R\left(\cos x,\ \sin x\right) dx.^{*}$$

If we make the substitution $z = e^{ix}$, the subject of integration becomes a rational function $H(z)$, which we may suppose split up into

(a)   a constant and certain positive and negative powers of z,

(b)   groups of terms of the type

$$\frac{A_{0}}{z - a} + \frac{A_{1}}{(z - a)^{2}} + \cdots + \frac{A_{n}}{(z - a)^{n+1}} \ldots\ldots\ldots\ldots\ldots\ldots.(1).$$

The terms (i), when expressed in terms of x, give rise to a term

$$\Sigma\left(c_{k} \cos kx + d_{k} \sin kx\right).$$

In the group (1) we put $z = e^{ix}$, $a = e^{i\alpha}$ and, using the equation

$$\frac{1}{z - a} = \tfrac{1}{2}e^{-i\alpha}\left\{-1 - i\cot\tfrac{1}{2}(x - a)\right\},$$

* See Hermite, *Cours d'analyse*, pp. 320 et seq.

we obtain a polynomial of degree $n+1$ in $\cot \frac{1}{2}(x-a)$. Since

$$\cot^2 x = -1 - \frac{d \cot x}{dx}, \qquad \cot^3 x = -\cot x - \frac{1}{2}\frac{d}{dx}(\cot^2 x), \, ...,$$

this polynomial may be transformed into the form

$$C + C_0 \cot \tfrac{1}{2}(x-a) + C_1 \frac{d}{dx}\cot \tfrac{1}{2}(x-a) + ... + C_n \frac{d^n}{dx^n}\cot \tfrac{1}{2}(x-a).$$

The function $R(\cos x, \sin x)$ is now expressed as a sum of a number of terms each of which is immediately integrable. The integral is a rational function of $\cos x$ and $\sin x$ if all the constants C_0 vanish; otherwise it includes a number of terms of the type

$$2C_0 \log \sin \tfrac{1}{2}(x-a).$$

Let us suppose for simplicity that $H(z)$, when split up into partial fractions, contains no terms of the types

$$C, \quad z^m, \quad z^{-m}, \quad (z-a)^{-p} \qquad (p > 1).$$

Then

$$R(\cos x, \sin x) = C_0 \cot \tfrac{1}{2}(x-a) + D_0 \cot \tfrac{1}{2}(x-\beta) + ...,$$

and the constants $C_0, D_0, ...$ may be determined by multiplying each side of the equation by $\sin \frac{1}{2}(x-a)$, $\sin \frac{1}{2}(x-\beta)$, ... and making x tend to $a, \beta, ...$.

It is often convenient to use the equation

$$\cot \tfrac{1}{2}(x-a) = \cot(x-a) + \operatorname{cosec}(x-a)$$

which enables us to decompose the function R into two parts $U(x)$ and $V(x)$ such that

$$U(x+\pi) = U(x), \quad V(x+\pi) = -V(x).$$

If R has the period π, then V must vanish identically; if it changes sign when x is increased by π, then U must vanish identically. Thus we find without difficulty that, if $m < n$,

$$\frac{\sin mx}{\sin nx} = \frac{1}{2n}\sum_0^{2n-1}\frac{(-1)^k \sin ma}{\sin(x-a)} = \frac{1}{n}\sum_0^{n-1}\frac{(-1)^k \sin ma}{\sin(x-a)},$$

or

$$\frac{\sin mx}{\sin nx} = \frac{1}{n}\sum_0^{n-1}(-1)^k \sin ma \cot(x-a),$$

where $a = k\pi/n$, according as $m+n$ is odd or even.

Similarly

$$\frac{1}{\sin(x-a)\sin(x-b)\sin(x-c)} = \Sigma\frac{1}{\sin(a-b)\sin(a-c)\sin(x-a)},$$

$$\frac{\sin(x-d)}{\sin(x-a)\sin(x-b)\sin(x-c)} = \Sigma\frac{\sin(a-d)}{\sin(a-b)\sin(a-c)}\cot(x-a).$$

(iii) One of the most important integrals in applications is

$$\int \frac{dx}{a+b\cos x},$$

where a and b are real. This integral may be evaluated in the manner explained above, or by the transformation $\tan \frac{1}{2}x = t$. A more elegant method is the following. If $|a| > |b|$, we suppose a positive, and use the transformation

$$(a + b \cos x)(a - b \cos y) = a^2 - b^2,$$

which leads to
$$\frac{dx}{a + b \cos x} = \frac{dy}{\sqrt{(a^2 - b^2)}}.$$

If $|a| < |b|$, we suppose b positive, and use the transformation

$$(b \cos x + a)(b \cosh y - a) = b^2 - a^2.$$

The integral
$$\int \frac{dx}{a + b \cos x + c \sin x}$$

may be reduced to this form by the substitution $x + a = y$, where $\cot a = b/c$. The forms of the integrals

$$\int \frac{dx}{(a + b \cos x)^n}, \quad \int \frac{dx}{(a + b \cos x + c \sin x)^n}$$

may be deduced by the use of formulae of reduction, or by differentiation with respect to a. The integral

$$\int \frac{dx}{(A \cos^2 x + 2B \cos x \sin x + C \sin^2 x)^n}$$

is really of the same type, since

$$A \cos^2 x + 2B \cos x \sin x + C \sin^2 x = \tfrac{1}{2}(A + C) + \tfrac{1}{2}(A - C) \cos 2x + B \sin 2x.$$

And similar methods may be applied to the corresponding integrals which contain hyperbolic functions, so that this type includes a large variety of integrals of common occurrence.

(iv) The same substitutions may of course be used when the subject of integration is an irrational function of $\cos x$ and $\sin x$, though sometimes it is better to use the substitutions $\cos x = t$, $\sin x = t$, or $\tan x = t$. Thus the integral

$$\int R(\cos x, \sin x, \sqrt{X}) \, dx,$$

where
$$X = (a, b, c, f, g, h \, \chi \cos x, \sin x, 1)^2,$$

is reduced to an elliptic integral by the substitution $\tan \tfrac{1}{2}x = t$. The most important integrals of this type are

$$\int \frac{R(\cos x, \sin x) \, dx}{\sqrt{(1 - k^2 \sin^2 x)}}, \quad \int \frac{R(\cos x, \sin x) \, dx}{\sqrt{(a + \beta \cos x + \gamma \sin x)}}.$$

3. The integral
$$\int P(x, e^{ax}, e^{bx}, \dots, e^{kx}) \, dx,$$

where $a, b, \dots, k$ are any numbers (commensurable or not), and P is a polynomial, is always an elementary function. For it is obvious that the integral can be reduced to the sum of a finite number of integrals of the type

$$\int x^p \, e^{Ax} \, dx \, ;$$

and
$$\int x^p \, e^{Ax} \, dx = \left(\frac{\partial}{\partial A}\right)^p \int e^{Ax} \, dx = \left(\frac{\partial}{\partial A}\right)^p \frac{e^{Ax}}{A} \, .$$

This type of integral includes a large variety of integrals, such as

$$\int x^m \, (\cos px)^\mu \, (\sin qx)^\nu \, dx, \quad \int x^m \, (\cosh px)^\mu \, (\sinh qx)^\nu \, dx,$$

$$\int x^m e^{-ax} \, (\cos px)^\mu \, dx, \quad \int x^m e^{-ax} \, (\sin qx)^\nu \, dx,$$

(m, μ, ν, being positive integers) for which formulae of reduction are given in text-books on the integral calculus.

Such integrals as

$$\int P \, (x, \, \log x) \, dx, \quad \int P \, (x, \, \text{arc} \sin x) \, dx, \, \ldots,$$

where P is a polynomial, may be reduced to particular cases of the above general integral by the obvious substitutions

$$x = e^y, \quad x = \sin y, \, \ldots.$$

4.   Except for the two classes of functions considered in the three preceding paragraphs, there are no really general classes of transcendental functions which we can *always* integrate in finite terms, although of course there are innumerable particular forms which may be integrated by particular devices.   There are however many classes of such integrals for which a systematic reduction theory may be given, analogous to the reduction theory for elliptic integrals. Such a reduction theory endeavours in each case

(i)   to split up any integral of the class under consideration into the sum of a number of parts of which some are elementary and the others not ;

(ii)   to reduce the number of the latter terms to the least possible ;

(iii)   to prove that these terms are incapable of further reduction, and are genuinely new and independent transcendents.

As an example of this process we shall consider the integral

$$\int e^x \, R \, (x) \, dx$$

where $R \, (x)$ is a rational function of x.*   The theory of partial

* See Hermite, *Cours d'analyse*, pp. 352 *et seq.*

fractions enables us to decompose this integral into the sum of a number of terms

$$A \int \frac{e^x}{x-a}\, dx, \quad A_m \int \frac{e^x}{(x-a)^{m+1}}\, dx, \cdots, \quad B \int \frac{e^x}{x-b}\, dx, \cdots.$$

Since

$$\int \frac{e^x}{(x-a)^{m+1}}\, dx = -\frac{e^x}{m\,(x-a)^m} + \frac{1}{m} \int \frac{e^x}{(x-a)^m}\, dx,$$

the integral may be further reduced so as to contain only

(i)   a term $\qquad\qquad e^x S(x)$

where $S(x)$ is a rational function ;

(ii)   a number of terms of the type

$$a \int \frac{e^x\, dx}{x-a}.$$

If all the constants a vanish, then the integral can be calculated in the finite form $e^x S(x)$. If they do not we can at any rate assert that the integral cannot be calculated *in this form**. For no such relation as

$$a \int \frac{e^x\, dx}{x-a} + \beta \int \frac{e^x\, dx}{x-b} + \ldots + \kappa \int \frac{e^x\, dx}{x-k} = e^x\, T(x),$$

where T is rational, can hold for all values of x. To see this it is only necessary to put $x = a + h$ and to expand in ascending powers of h. Then

$$a \int \frac{e^x\, dx}{x-a} = a e^a \int \frac{e^h}{h}\, dh$$

$$= a e^a (\log h + h + \ldots),$$

and no *logarithm* can occur in any of the other terms†.

Consider, for example, the integral

$$\int e^x \left(1-\frac{1}{x}\right)^3 dx.$$

This is equal to $\qquad e^x - 3 \int \frac{e^x}{x}\, dx + 3 \int \frac{e^x}{x^2}\, dx - \int \frac{e^x}{x^3}\, dx,$

and since $\qquad 3 \int \frac{e^x}{x^2}\, dx = -\frac{3e^x}{x} + 3 \int \frac{e^x}{x}\, dx,$

and

$$-\int \frac{e^x}{x^3}\, dx = \frac{e^x}{2x^2} - \frac{1}{2} \int \frac{e^x}{x^2}\, dx = \frac{e^x}{2x^2} + \frac{e^x}{2x} - \frac{1}{2} \int \frac{e^x}{x}\, dx,$$

---

* See the remarks at the end of this paragraph.

† It is not difficult to give a purely algebraical proof on the lines of IV., § 2.

we obtain finally

$$\int e^x \left(1 - \frac{1}{x}\right)^3 dx = e^x \left(1 - \frac{7}{2x} + \frac{1}{2x^2}\right) - \frac{1}{2}\int \frac{e^x}{x}\, dx.$$

Similarly it will be found that

$$\int e^x \left(1 - \frac{2}{x}\right)^2 dx = 2e^x \left(\frac{1}{2} - \frac{2}{x}\right),$$

this integral being an elementary function.

Since
$$\int \frac{e^x}{x - a}\, dx = e^a \int \frac{e^y}{y}\, dy,$$

if $x = y + a$, all integrals of this kind may be made to depend on known functions and on the single transcendent

$$\int \frac{e^x}{x}\, dx,$$

which is usually denoted by $Li\, e^x$ and is of great importance in the theory of numbers. The question of course arises as to whether this integral is not itself an elementary function.

Now Liouville* has proved the following theorem: '*if y is any algebraical function of x, and*

$$\int e^x y\, dx$$

is an elementary function, then

$$\int e^x y\, dx = e^x \left(\alpha + \beta y + \ldots + \lambda y^{n-1}\right),$$

$\alpha, \beta, \ldots, \lambda$ *being rational functions of x and n the degree of the algebraical equation which determines y as a function of x*'.

Liouville's proof rests on the same general principles as do those of the corresponding theorems concerning the integral $\int y\, dx$. It will be observed that no logarithmic terms can occur, and that the theorem is therefore very similar to that which holds for $\int y\, dx$ in the simple case in which the integral is *algebraical*. The argument which shows that no logarithmic terms occur is substantially the same as that which shows that, when they occur in the integral of an algebraical function, they must occur linearly. In this case the occurrence of the exponential factor precludes even this possibility, since differentiation will not eliminate logarithms when they occur in the form

$$e^x \log f(x).$$

* ' Mémoire sur l'intégration d'une classe de fonctions transcendantes ', *Journal für Mathematik*, vol. 13, 1835, pp. 93–118. Liouville shows how the integral, when of this form, may always be calculated by elementary methods.

In particular, if y is a rational function, then the integral must be of the form

$$e^x R(x)$$

and this we have already seen to be impossible. Hence the 'logarithm-integral'

$$Li\ e^x = \int \frac{e^x}{x}\,dx = \int^{e^x} \frac{dy}{\log y}$$

is really a new transcendent, which cannot be expressed in finite terms by means of elementary functions ; and the same is true of all integrals of the type

$$\int e^x R(x)\,dx$$

which cannot be calculated in finite terms by means of the process of reduction sketched above.

The integrals

$$\int \sin x\ R(x)\,dx, \quad \int \cos x\ R(x)\,dx$$

may be treated in a similar manner. Either the integral is of the form

$$\cos x\ R_1(x) + \sin x\ R_2(x)$$

or it consists of a term of this kind together with a number of terms which involve the transcendents

$$\int \frac{\cos x}{x}\,dx, \quad \int \frac{\sin x}{x}\,dx,$$

which are called the cosine-integral and sine-integral of x, and denoted by $Ci\ x$ and $Si\ x$. These transcendents are of course not fundamentally distinct from the logarithm-integral.

5. Liouville has gone further and shown that it is always possible to determine whether the integral

$$\int (Pe^p + Qe^q + \ldots + Te^t)\,dx,$$

where $P, Q, \ldots, T, p, q, \ldots, t$ are algebraical functions, is an elementary function, and to obtain the integral in case it is one*. The most general theorem which has been proved in this region of mathematics, and which is also due to Liouville, is the following.

---

* An interesting particular result is that the 'error function' $\int e^{-x^2}\,dx$ is not an elementary function.

'*If y, z, ... are functions of x whose differential coefficients are algebraical functions of x, y, z, ..., and F denotes an algebraical function, and if*

$$\int F(x, y, z, ...)\,dx$$

is an elementary function, then it is of the form

$$t + A\log u + B\log v + ...,$$

where t, u, v, ... are algebraical functions of x, y, z, ... If the differential coefficients are rational in x, y, z, ..., and F is rational, then t, u, v, ... are rational in x, y, z, ...'

Thus for example the theorem applies to

$$F(x, e^x, e^{e^x}, \log x, \log\log x, \cos x, \sin x),$$

since, if the various arguments of F are denoted by x, y, z, ξ, η, ζ, θ, we have

$$\frac{dy}{dx} = y, \qquad \frac{dz}{dx} = yz, \qquad \frac{d\xi}{dx} = \frac{1}{x},$$

$$\frac{d\eta}{dx} = \frac{1}{x\xi}, \qquad \frac{d\zeta}{dx} = -\sqrt{(1-\zeta^2)}, \qquad \frac{d\theta}{dx} = \sqrt{(1-\theta^2)}.$$

The proof of the theorem does not involve ideas different in principle from those which have been employed continually throughout the preceding pages.

6. As a final example of the manner in which these ideas may be applied, we shall consider the following question :

'*in what circumstances is*

$$\int R(x)\log x\,dx,$$

where R is rational, an elementary function ?'

In the first place the integral must be of the form

$$R_0(x, \log x) + A_1\log R_1(x, \log x) + A_2\log R_2(x, \log x) + ...$$

A general consideration of the form of the differential coefficient of this expression, in which $\log x$ must only occur linearly and multiplied by a rational function, leads us to anticipate that (i) $R_0(x, \log x)$ must be of the form

$$S(x)(\log x)^2 + T(x)\log x + U(x),$$

where S, T, and U are rational, and (ii) R_1, R_2, ... must be rational functions of x only ; so that the integral can be expressed in the form

$$S(x)(\log x)^2 + T(x)\log x + U(x) + \Sigma B_k\log(x - a_k).$$

Differentiating, and comparing the result with the subject of integration, we obtain the equations

$$S'=0, \quad \frac{2S}{x}+T'=R, \quad \frac{T}{x}+U'+\Sigma\frac{B_k}{x-a_k}=0.$$

Hence S is a constant, say $\tfrac{1}{2}C$, and

$$T=\int\left(R-\frac{C}{x}\right)dx.$$

We can always determine by means of elementary operations, as in IV., § 4, whether this integral is rational for any value of C or not. If not, then the given integral is not an elementary function. If T is rational, then we must calculate its value, and substitute it in the integral

$$U=-\int\left\{\frac{T}{x}+\Sigma\frac{B_k}{x-a_k}\right\}dx=-\int\frac{T}{x}dx-\Sigma B_k\log{(x-a_k)},$$

which must be rational for some value of the arbitrary constant implied in T. We can calculate the rational part of

$$\int\frac{T}{x}dx:$$

the transcendental part must be cancelled by the logarithmic terms

$$\Sigma B_k\log{(x-a_k)}.$$

The necessary and sufficient condition that the original integral should be an elementary function is therefore that R should be of the form

$$\frac{C}{x}+\frac{d}{dx}\{R_1(x)\},$$

where C is a constant and R_1 is rational. That the integral is in this case such a function becomes obvious if we integrate by parts, for

$$\int\left(\frac{C}{x}+R_1'\right)\log x\,dx=\tfrac{1}{2}C(\log x)^2+R_1\log x-\int\frac{R_1}{x}dx.$$

In particular

(i) $\displaystyle\int\frac{\log x}{x-a}dx,$ \qquad (ii) $\displaystyle\int\frac{\log x}{(x-a)(x-b)}dx,$

are not elementary functions unless in (i) $a=0$ and in (ii) $b=a$. If the integral is elementary then the integration can always be carried out, with the same reservation as was necessary in the case of rational functions.

It is evident that the problem considered in this paragraph is but one of a whole class of similar problems. The reader will find it instructive to formulate and consider such problems for himself.

7. It will be obvious by now that the number of classes of transcendental functions whose integrals are always elementary is very small, and that such integrals as

$$\int f(x, e^x)\, dx, \qquad\qquad \int f(x, \log x)\, dx,$$

$$\int f(x, \cos x, \sin x)\, dx. \quad \int f(e^x, \cos x, \sin x)\, dx,$$

$$\dots\dots\dots\dots\dots ,$$

where f is algebraical, or even rational, are generally new transcendents. These new transcendents, like the transcendents (such as the elliptic integrals) which arise from the integration of algebraical functions, are in many cases of great interest and importance. They may often be expressed by means of infinite series or definite integrals, or their properties may be studied by means of the integral expressions which define them. The very fact that such a function is *not* an elementary function in so far enhances its importance. And when such functions have been introduced into analysis new problems of integration arise in connection with them. We may enquire, for example, under what circumstances an elliptic integral or elliptic function, or a combination of such functions with elementary functions, can be integrated in finite terms by means of elementary and elliptic functions. But before we can be in a position to restate the fundamental problem of the Integral Calculus in any such more general form, it is essential that we should have disposed of the particular problem formulated in Section III.

APPENDIX I

BIBLIOGRAPHY

The following is a list of the memoirs by Abel, Liouville and Tschebyschef which have reference to the subject matter of this tract.

N. H. Abel

1. 'Über die Integration der Differential-Formel $\frac{\rho dx}{\sqrt{R}}$, wenn R und ρ ganze Funktionen sind', *Journal für Mathematik*, vol. 1, 1826, pp. 185–221 (*Œuvres*, vol. 1, pp. 104–144).

2. 'Précis d'une théorie des fonctions elliptiques', *Journal für Mathematik*, vol. 4, 1829, pp. 236–277, 309–348 (*Œuvres*, vol. 1, pp. 518–617).

3. 'Théorie des transcendantes elliptiques', *Œuvres*, vol. 2, pp. 87–188.

J. Liouville

1. 'Mémoire sur la classification des transcendantes, et sur l'impossibilité d'exprimer les racines de certaines équations en fonction finie explicite des coefficients', *Journal de mathématiques*, ser. 1, vol. 2, 1837, pp. 56–104.

2. 'Nouvelles recherches sur la détermination des intégrales dont la valeur est algébrique', *ibid.*, vol. 3, 1838, pp. 20–24 (previously published in the *Comptes Rendus*, 28 Aug. 1837).

3. 'Suite du mémoire sur la classification des transcendantes, et sur l'impossibilité d'exprimer les racines de certaines équations en fonction finie explicite des coefficients', *ibid.*, pp. 523–546

4. 'Note sur les transcendantes elliptiques considérées comme fonctions de leur module', *ibid.*, vol. 5, 1840, pp. 34–37.

5. 'Mémoire sur les transcendantes elliptiques considérées comme fonctions de leur module', *ibid.*, pp. 441–464.

6. 'Premier mémoire sur la détermination des intégrales dont la valeur est algébrique', *Journal de l'École Polytechnique*, vol. 14, cahier 22, 1833, pp. 124–148 (also published in the *Mémoires présentés par divers savants à l'Académie des Sciences*, vol. 5, 1838, pp. 76–151).

7. 'Second mémoire sur la détermination des intégrales dont la valeur est algébrique', *ibid.*, pp. 149–193 (also published as above).

8. ‘ Mémoire sur les transcendantes elliptiques considérées comme fonctions de leur amplitude ’, *ibid.*, cahier 23, 1834, pp. 37–83.

9. ‘ Mémoire sur l’intégration d’une classe de fonctions transcendantes ’, *Journal für Mathematik*, vol. 13, 1835, pp. 93–118.

P. Tschebyschef

1. ‘ Sur l’intégration des différentielles irrationnelles ’, *Journal de mathématiques*, ser. 1, vol. 18, 1853, pp. 87–111 (*Œuvres*, vol. 1, pp. 147–168).

2. ‘ Sur l’intégration des différentielles qui contiennent une racine carrée d’une polynome du troisième ou du quatrième degré ’, *ibid.*, ser. 2, vol. 2, 1857, pp. 1–42 (*Œuvres*, vol. 1, pp. 171–200 ; also published in the *Mémoires de l’Académie Impériale des Sciences de St-Pétersbourg*, ser. 6, vol. 6, 1857, pp. 203–232).

3. ‘ Sur l’intégration de la différentielle $\dfrac{x+A}{\sqrt{(x^4+ax^3+\beta x^2+\gamma x+\delta)}}\,dx$ ’, *ibid.*, ser. 2, vol. 9, 1864, pp. 225–241 (*Œuvres*, vol. 1, pp. 517–530 ; previously published in the *Bulletin de l’Académie Impériale des Sciences de St-Pétersbourg*, vol. 3, 1861, pp. 1–12).

4. ‘ Sur l’intégration des différentielles irrationnelles ’, *ibid.*, pp. 242–246 (*Œuvres*, vol. 1, pp. 511–514 ; previously published in the *Comptes Rendus*, 9 July 1860).

5. ‘ Sur l’intégration des différentielles qui contiennent une racine cubique ’ (*Œuvres*, vol. 1, pp. 563–608 ; previously published only in Russian).

Other memoirs which may be consulted are :

A. Clebsch

‘ Über diejenigen Curven, deren Coordinaten sich als elliptische Functionen eines Parameters darstellen lassen ’, *Journal für Mathematik*, vol. 64, 1865, pp. 210–270.

J. Dolbnia

‘ Sur les intégrales pseudo-elliptiques d’Abel ’, *Journal de mathématiques*, ser. 4, vol. 6, 1890, pp. 293–311.

Sir A. G. Greenhill

‘ Pseudo-elliptic integrals and their dynamical applications ’, *Proc. London Math. Soc.*, ser. 1, vol. 25, 1894, pp. 195–304.

G. H. Hardy

‘ Properties of logarithmico-exponential functions ’, *Proc. London Math. Soc.*, ser. 2, vol. 10, 1910, pp. 54–90.

L. Königsberger

‘ Bemerkungen zu Liouville’s Classificirung der Transcendenten ’, *Mathematische Annalen*, vol. 28, 1886, pp. 483–492.

L. Raffy

'Sur les quadratures algébriques et logarithmiques', *Annales de l'École Normale*, ser. 3, vol. 2, 1885, pp. 185–206.

K. Weierstrass

'Über die Integration algebraischer Differentiale vermittelst Logarithmen', *Monatsberichte der Akademie der Wissenschaften zu Berlin*, 1857, pp. 148–157 (*Werke*, vol. 1, pp. 227–232).

G. Zolotareff

'Sur la méthode d'intégration de M. Tschebyschef', *Journal de mathématiques*, ser. 2, vol. 19, 1874, pp. 161–188.

Further information concerning pseudo-elliptic integrals, and degenerate cases of Abelian integrals generally, will be found in a number of short notes by Dolbnia, Kapteyn and Ptaszycki in the *Bulletin des sciences mathématiques*, and by Goursat, Gunther, Picard, Poincaré, and Raffy in the *Bulletin de la Société Mathématique de France*, in Legendre's *Traité des fonctions elliptiques* (vol. 1, ch. 26), in Halphen's *Traité des fonctions elliptiques* (vol. 2, ch. 14), and in Enneper's *Elliptische Funktionen*. The literature concerning the general theory of algebraical functions and their integrals is too extensive to be summarised here: the reader may be referred to Appell and Goursat's *Théorie des fonctions algébriques*, and Wirtinger's article *Algebraische Funktionen und ihre Integrale* in the *Encyclopädie der Mathematischen Wissenschaften*, II B 2.

66

APPENDIX II

ON ABEL'S PROOF OF THE THEOREM OF V., § 11

Abel's proof (*Œuvres*, vol. 1, p. 545) is as follows*:
We have

$$\psi\,(x,\,u)=0 \quad\dots\dots\dots\dots\dots\dots\dots\dots(1),$$

where ψ is an irreducible polynomial of degree m in u. If we make use of the equation $f(x,\,y)=0$, we can introduce y into this equation, and write it in the form

$$\phi\,(x,\,y,\,u)=0 \quad\dots\dots\dots\dots\dots\dots\dots(2),$$

where ϕ is a polynomial in the three variables x, y, and u†; and we can suppose ϕ, like ψ, of degree m in u and irreducible, that is to say not divisible by any polynomial of the same form which is not a constant multiple of ϕ or itself a constant.

From $f=0$, $\phi=0$ we deduce

$$\frac{\partial f}{\partial x}+\frac{\partial f}{\partial y}\frac{dy}{dx}=0,\qquad \frac{\partial\phi}{\partial x}+\frac{\partial\phi}{\partial y}\frac{dy}{dx}+\frac{\partial\phi}{\partial u}\frac{du}{dx}=0\,;$$

and, eliminating $\dfrac{dy}{dx}$, we obtain an equation of the form

$$\frac{du}{dx}=\frac{\lambda\,(x,\,y,\,u)}{\mu\,(x,\,y,\,u)},$$

where λ and μ are polynomials in x, y, and u. And in order that u should be an integral of y it is necessary and sufficient that

$$\lambda-y\mu=0 \quad\dots\dots\dots\dots\dots\dots\dots(3).$$

Abel now applies Lemma (2) of § 11, or rather its analogue for polynomials in u whose coefficients are polynomials in x and y, to the two polynomials ϕ and $\lambda-y\mu$, and infers that *all* the roots $u,\,u',\dots$ of $\phi=0$ satisfy (3). From this he deduces that $u,\,u',\dots$ are all integrals of y, and so that

$$\frac{u+u'+\dots}{m+1} \quad\dots\dots\dots\dots\dots\dots\dots(4)$$

* The theorem with which Abel is engaged is a very much more general theorem.

† 'Or, au lieu de supposer ces coefficiens rationnels en x, nous les supposerons rationnels en x, y; *car cette supposition permise simplifiera beaucoup le raisonnement*'.

is an integral of y.   As (4) is a symmetric function of the roots of (2), it is a rational function of x and y, whence his conclusion follows *.

It will be observed that the hypothesis that (2) does actually involve y is essential, if we are to avoid the absurd conclusion that u is necessarily *a rational function of x only*.   On the other hand it is not obvious how the presence of y in ϕ affects the other steps in the argument.

The crucial inference is that which asserts that because the equations $\phi = 0$ and $\lambda - y\mu = 0$, considered as equations in u, have a root in common, and ϕ is irreducible, therefore $\lambda - y\mu$ is divisible by ϕ.   This inference is invalid.

We could only apply the lemma in this way if the equation (3) were satisfied by one of the roots of (2) *identically*, that is to say for all values of x and y.   But this is not the case.   The equations are satisfied by the same value of u *only when x and y are connected by the equation* (1).

Suppose, for example, that

$$y = \frac{1}{\sqrt{(1+x)}}, \quad u = 2\sqrt{(1+x)}.$$

Then we may take

$$f = (1+x)\,y^2 - 1,$$

$$\psi = u^2 - 4\,(1+x),$$

and

$$\phi = uy - 2.$$

Differentiating the equations $f = 0$ and $\phi = 0$, and eliminating $\dfrac{dy}{dx}$, we find

$$\frac{du}{dx} = \frac{u}{2\,(1+x)} = \frac{\lambda}{\mu}.$$

Thus

$$\phi = uy - 2, \quad \lambda - y\mu = u - 2y\,(1+x)\,;$$

and these polynomials have a common factor only in virtue of the equation $f = 0$.

---

* Bertrand (*Calcul intégral*, ch. 5) replaces the last step in Abel's argument by the observation that if u and u' are both integrals of y then $u - u'$ is constant (cf. p. 39, bottom).   It follows that the degree of the equation which defines u can be decreased, which contradicts the hypothesis that it is irreducible.